GNVQ Core Skills

APPLICATION OF NUMBER

Intermediate/Advanced

GNVQ Core Skills

APPLICATION OF NUMBER

Intermediate/Advanced

Catriona Johnstone

Lecturer in Application of Number and a GNVQ Course Manager
Farnborough College of Technology

PITMAN PUBLISHING

PITMAN PUBLISHING
128 Long Acre, London WC2E 9AN

A Division of Longman Group Limited

First published in1994

A CIP catalogue record for this book can be obtained from the British Library.

ISBN 0 273 60625 5

Typeset by M Rules
Printed and bound in Great Britain by Clays Ltd, St Ives plc

The Publishers' policy is to use paper manufactured from sustainable forests.

To Newton, Owen and Blaine

Contents

Introduction

The Application of Number core skill unit identifies three key themes:

- Gathering and processing data (Element 1)
- Representing and tackling problems (Element 2)
- Interpreting and presenting data (Element 3)

Both Intermediate and Advanced candidates are required to fulfil the same performance criteria. Both are also required to employ the same mathematical techniques (with a few exceptions at Intermediate Level). The main difference between the two levels is that Advanced Level candidates will perform more complex mathematics than Intermediate Level candidates. The book does not distinguish between questions for Intermediate candidates and those for Advanced candidates. It is the responsibility of the candidate's assessor to decide whether the evidence submitted is at Level 2 or Level 3. However, the questions in the book do get progressively more difficult.

The National Council for Vocational Qualifications (NCVQ) took the mathematical techniques covered in the Application of Number core skill unit from National Curriculum statements of attainment. NCVQ used only those statements of attainment that they deemed to be relevant to a wide range of vocational settings. For cross-reference purposes, the National Curriculum reference numbers are included in the text.

The aim of GNVQ Application of Number is to help the candidate understand and apply, in a vocational setting, much of the mathematics learned at school. For many, this book will only be used for revision and as a guide to what is required as evidence. For others, it will be an essential study aid to learn the mathematical techniques required to obtain a GNVQ qualification.

Before you begin learning the mathematical techniques, you should understand how the system works. The Application of Number core skill unit consists of three elements. Each element has its own performance criteria and range. The range contains the mathematical techniques you need to demonstrate (the mathematics questions you need to answer correctly). To pass each mathematical technique you must fulfil all the performance criteria. For example, Element 1 at Intermediate Level has five mathematical techniques and six performance criteria. Each mathematical technique must be employed at least once, but each performance criteria will have to be fulfilled at least five times (once for every mathematical technique).

The structure of this book is such that each part covers one complete element, and each chapter covers one technique in the range for that element. Examples of how the demonstration of the techniques would fulfil the performance criteria are given in each chapter.

Element 1: Gather and process data

Performance criteria

1 Techniques appropriate to the task are selected and used by the individual.
2 Activities required by the techniques are performed to appropriate levels of precision and in the correct sequence.
3 Mathematical terms and mathematical aspects of everyday language are used and interpreted with precision (AT1 3b).
4 Data are recorded in the required units.
5 Data are recorded in the required format.
6 Records are accurate and complete.

Range

Techniques

- Make estimates based on familiar units of measurement, checking results (AT2 3d).
- Convert between different units of measurement using tables, graphs and scales.
- Conduct a survey on an issue of the individual's choice (AT5 4b).
- Design and use an observation sheet to collect data (AT5 5b).
- Design and use a questionnaire to survey opinion (AT5 6a).

Advanced level candidates only:

- Organise data into groups and classifications.

Element 2: Represent and tackle problems

Performance criteria

1 Techniques appropriate to the task are selected and used by the individual.
2 Activities required by the techniques are performed to appropriate levels of precision and in the correct sequence.
3 Mathematical terms and mathematical aspects of everyday language are used and interpreted with precision (AT1 3b).
4 Use estimation to check calculations (AT2 4c).
5 Calculations and results obtained are correct.
6 Aids are used correctly.
7 Conclusions/generalisations/predictions drawn from results are valid.
8 Clear justification is given for the interpretation of results.

Range

Techniques

- Solve whole number problems involving addition and subtraction (AT2 2b).
- Solve problems involving multiplication and division (AT2 3c).
- Use fractions, decimals and percentages to describe situations (AT2 4c).
- Calculate with fractions, decimals, percentages and ratios (AT2 6a).
- Find perimeters, areas and volumes (AT4 4d).
- Find areas of plane shapes and volumes of simple solids (AT4 5d).
- Know and use the formulae for finding the area and circumference of circles (AT4 6d).
- Use simple formulae expressed in words (AT3 4b).
- Solve simple equations (AT3 6b).
- Use networks to solve problems (AT4 5c).

Advanced level candidates only:

- Carry out calculations in plane and solid shapes (AT4 7d).
- Solve equations or simple inequalities (AT3 7b).
- Use symbolic notation to express the rules of sequences (AT3 7a).

■ Element 3: Interpret and present mathematical data

Performance criteria

1 Trends/significant features of data are identified correctly.
2 Conclusions/generalisations/predictions drawn from the data are valid.
3 Mathematical terms and mathematical aspects of everyday language are used and interpreted with precision (AT1 3b).
4 Clear justification is given for the interpretation of data.
5 Conventions for presenting data are followed accurately.
6 Data are represented accurately and to appropriate levels of precision.

Range

Data

Produced by working through activities described in elements 1 and 2.

Techniques

- Use mathematical terms to describe common 2D shapes and 3D objects (AT4 2a).
- Use 2D representation of 3D objects (AT4 6a).
- Construct and interpret statistical diagrams (AT5 3b).
- Use symbols and diagrams.
- Use the mean and range of a set of data (AT5 4c).
- Identify all the outcomes of combining two independent events (AT5 6c).

Advanced level candidates only:

- Given the probability of exclusive events, calculate the probability of a combined event (AT5 7c).

Performance criteria

Before attempting to master the mathematical techniques, you should make sure you understand what is required to cover the performance criteria.

Techniques appropriate to the task are selected and used by the individual.

This means that you must choose and use the correct mathematical technique for a particular task. For example, you would not conduct a survey to find out the present population of the UK; you would look at a statistical table or graph.

Activities required by the techniques are performed to appropriate levels of precision and in the correct sequence.

If you were conducting a survey of the number of working and broken light bulbs in a building, your correct sequence of activities would be:

- Compile a list of all the rooms with numbers or position in the building.
- Add a column next to the list for a tally of all the bulbs working and a column for a tally of all the bulbs broken.
- Visit all the rooms on the list, completing the tally as you go from room to room.

The appropriate level of precision for this task would be to visit all the rooms and check all the bulbs in the building.

Mathematical terms and mathematical aspects of everyday language are used and interpreted with precision (AT1 3b).

Fraction is a mathematical term. Let us say that you have surveyed 20 people regarding their opinion on smoking in the workplace. In your analysis you state that one half (50%) thinks smoking should be banned in the workplace. You must have evidence that ten people have responded as indicated. If however only eight of the people questioned responded as indicated, then you must state that two fifths (40%) think smoking should be banned in the workplace.

Data are recorded in the required units.

Here you must remember to insert £ signs, cm, ft, lt, kg, °C, etc. when calculating in any of these or other units.

Data are recorded in the required format.

Some examples of required formats are:

- Numbers
- Words
- Letters
- 𝍸 on a tally chart used to conduct a survey
- Words, letters or numbers on a two-way chart used to organise data into groups and classifications.

Records are accurate and complete.

You must have no calculation mistakes or missing answers.

Use estimation to check calculations (AT2 4c).

You must show the use of estimation, either by calculating on paper or working a calculation out in your head and then writing the answer down.

Calculations and results obtained are correct.

Both the final result and all calculations used to get that result must be correct.

Aids are used correctly.

Some examples of aids are:

- calculator
- ruler
- compass
- protractor
- set square
- graph paper
- logarithms
- isometric grid paper

On many occasions, the fact that you have the correct answer demonstrates that you can use the aid correctly. However, this is not always evident. For this reason your tutor may wish to observe your use of the aid.

Conclusions/generalisations/predictions drawn from results/data are valid.

If your data indicates:

> Each year sees an increase in the number of cars on the roads of Britain.

The conclusions/generalisations drawn from this could be:

- The driving test is easier to pass.
- Cars are cheaper to buy.
- More people have the money to buy cars.
- Public transport is more expensive.
- It is fashionable to drive a car.
- More people own their own home.

Some of these conclusions/generalisations are valid and some are not. For example, the fact that more people own their own home has no direct effect on the number of cars on the roads, but the cost of cars and the cost of public transport have.

Clear justification is given for the interpretation of results/data.

If, in the previous performance criterion, you concluded that there are more cars on the roads of Britain because the cost of public transport is more expensive, you would need to justify this statement with a table of figures (e.g. the average cost of a ten-mile bus journey over the past five years).

Trends/significant features of data are identified correctly.

Trends are commonly found in sales figures, especially seasonal products like ice cream. You must not make mistakes when analysing data.

Conventions for presenting data are followed accurately.

For example, the conventions on a graph are: main title, axis titles, constant scale, etc.

Data are represented accurately and to appropriate levels of precision.

You must not make mistakes when drawing graphs, diagrams, symbols, etc.

The appropriate level of precision on a graph would be to use a sharp pencil and to join the points accurately.

Part 1
GATHER AND PROCESS DATA

In any vocation there is a constant need to gather data and process it correctly. The large corporation needs to know how well its product is doing in the market place; the local recreation centre needs to know which courses are popular, the residential elderly persons' home needs to know what medications will be required for the month. All this information can be collected and processed effectively by following set mathematical techniques. This section introduces you to those mathematical techniques.

1 Make estimates based on familiar units of measurement, checking results

Estimation is something you will use very often in your chosen vocation. You may need to estimate the amount of room required for a filing cabinet or desk, the length of a car, the amount of paint needed to paint a sign, the amount of time left to complete a task.

It is also important that you get into the habit of estimating your answers when using a calculator. It is very easy to press a wrong key on a calculator, thus obtaining an incorrect answer. If you have estimated, however, you will know the answer is wrong.

Metric and imperial measures

In the United Kingdom two types of measurement are in use: metric and imperial measures. In 1963 an act of Parliament specified that the imperial system of weights and measures be replaced with the metric system. This change has taken a long time, and is still not complete. It is only in recent years that petrol filling stations have converted their pumps to litres from gallons. However, road signs still indicate distance in miles. Industry now measures and weighs using the metric system, but there are many people in industry who still use the imperial system.

You should be so familiar with measurement that you can estimate length, weight and capacity in both metric and imperial measures.

Study Tables 1.1 and 1.2 and using a tape measure, a set of scales and a measuring jug, familiarise yourself with metric and imperial measures.

Table 1.1 Metric measures

Length	millimetre
	centimetre
	decimetre
	metre
	kilometre
Weight	gram
	kilogram
	tonne
Capacity	millilitre
	centilitre
	litre

Table 1.2 Imperial measures

Length	inch
	foot
	yard
	mile
Weight	ounce
	pound
	stone
Capacity	fluid ounce
	pint
	gallon

You will only learn these measures by experience, so start practising today.

Exercise 1.1

1 Estimate the lengths of the following items in centimetres and inches:

(a) a desk
(b) a book
(c) a camera
(d) a microwave oven
(e) a tin of beans.

Now check your results by using a ruler or tape measure.

2 Estimate the lengths of the following items in metres and feet:

(a) a car
(b) a corridor
(c) a room
(d) a garden fence
(e) a playing field.

Now check your results by using a ruler or tape measure.

3 Estimate the weights of the following items in grams and ounces:

(a) a personal stereo
(b) a book
(c) a child's toy
(d) a hammer
(e) a camera.

Now check your results by using a set of kitchen scales.

4 Estimate the weights of the following items in kilograms and pounds:

(a) a chair
(b) a vacuum cleaner
(c) a member of your family
(d) a garden spade
(e) a basket of laundry.

Now check your results by using a set of bathroom scales.

Once you become familiar with length, weight and capacity you should move on to learn how many smaller units make up the larger units. Tables 1.3 and 1.4 will be of help.

Table 1.3 Metric measures

Length	10 millimetres	(mm)	=	1 centimetre	(cm)
	1000 millimetres	(mm)	=	1 metre	(m)
	10 centimetres	(cm)	=	1 decimetre	(dm)
	10 decimetres	(dm)	=	1 metre	(m)
	100 centimetres	(cm)	=	1 metre	(m)
	1000 metres	(m)	=	1 kilometre	(km)
Weight	1000 milligrams	(mg)	=	1 gram	(g)
	1000 grams	(g)	=	1 kilogram	(kg)
	1000 kilograms	(kg)	=	1 tonne	(t)
Capacity	10 millilitres	(ml)	=	1 centilitre	(cl)
	1000 millilitres	(ml)	=	1 litre	(l)
	100 centilitres	(cl)	=	1 litre	(l)

Table 1.4 Imperial measures

Length	12 inches	(in)	=	1 foot	(ft)
	36 inches	(in)	=	1 yard	(yd)
	3 feet	(ft)	=	1 yard	(yd)
	1760 yards	(yd)	=	1 mile	
Weight	16 ounces	(oz)	=	1 pound	(lb)
	14 pounds	(lb)	=	1 stone	(st)
Capacity	20 fluid ounces	(fl oz)	=	1 pint	(pt)
	8 pints	(pt)	=	1 gallon	(gal)

EXAMPLE 1

(a) Convert 40 centimetres into millimetres.

$40 \times 10 = 400$

Answer = 400 mm

(b) How many yards are there in 3 miles?

$3 \times 1760 = 5280$

Answer = 5280 yards

(c) Convert 3,400 grams into kilograms.

$3400 \div 1000 = 3.4$

Answer = 3.4 kg

(d) Convert 24 inches into feet.

$24 \div 12 = 2$

Answer = 2 feet

(e) Convert 5 litres into millilitres.

$5 \times 1000 = 5000$

Answer = 5000 ml

(f) Convert 60 fluid ounces into pints.

$60 \div 20 = 3$

Answer = 3 pints

Figure 1.1 shows the performance criteria analysis for this example.

	Performance Criteria	Pass or NYA
1	techniques appropriate to the task are selected and used by the individual	Pass. Used multiplication or division
2	activities required by the techniques are performed to appropriate levels of precision and in the correct sequence	Pass. took original data and × or ÷ with conversion equivalent
3	mathematical terms and mathematical aspects of everyday language are used and interpreted with precision (AT1 3b)	Pass. Used the verb "to convert"
4	data are recorded in the required units	Pass. Used mm, yds, kg, etc
5	data are recorded in the required format	Pass used numbers
6	records are accurate and complete	Pass all Correct

Fig 1.1

Exercise 1.2

1 Convert the following to centimetres:

(a) 5 m (b) 30 mm (c) 270 mm (d) 4.27 m

2 Convert the following to millimetres:

(a) 21 cm (b) 315 cm (c) 4.6 cm (d) 5 m

3 Convert the following to feet:

(a) 3 yd (b) 48 in (c) $2\frac{2}{3}$ yd (d) 30 in

4 Convert the following to yards:

(a) 72 in (b) 12 ft (c) 2 miles (d) 108 in

5 Convert the following to metres:

(a) 500 cm (b) 3 km (c) 18 000 mm (d) 4.87 km

6 Convert the following to kilometres:

(a) 17 000 m (b) 500 000 cm (c) 6400 m (d) 6 430 000 cm

7 Convert the following to grams:

(a) 2000 mg (b) 4 kg (c) 8300 mg (d) 17.54 kg

8 Convert the following to kilograms:

(a) 7000 g (b) 25 000 000 mg (c) 3 t (d) 670 g

9 Convert the following to ounces:

(a) 2 lb (b) 3 st (c) 4 lb 7 oz (d) 1 st 8 lb

10 Convert the following to pounds:

(a) 64 oz (b) 5 st (c) 48 oz (d) 2 st 2 lb

11 Convert the following to millilitres:

(a) 2 l (b) 30 cl (c) 1.45 l (d) 74 cl

12 Convert the following to litres:

(a) 5000 ml (b) 200 cl (c) 3560 ml (d) 490 cl

13 Convert the following to pints:

(a) 40 fl oz (b) 8 gals (c) 50 fl oz (d) $4\frac{1}{8}$ gals

14 Convert the following to gallons:

(a) 16 pt (b) 480 fl oz (c) 44 pt (d) 720 fl oz

Measurement of time

Time is also a unit of measurement you are required to be familiar with.

1 second (sec)	=	a very short space of time
60 seconds	=	1 minute (min)
60 minutes	=	1 hour (hr)
24 hours	=	1 day

EXAMPLE 2

(a) Add 1 hour 23 minutes to 3 hours 7 minutes.

23 + 7 = 30 min
1 + 3 = 4 hr

Answer = 4 hr 30 min

(b) Subtract 1 hour 5 minutes from 4 hours 18 minutes.

18 – 5 = 13 min
4 – 1 = 3 hr

Answer = 3 hr 13 min

(c) Add 2 hr 37 min to 3 hr 55 min.

37 + 55 = 92 min = 1 hr 32 min
2 + 3 + 1 = 6 hr

Answer = 6 hr 32 min

Figure 1.2 gives the performance criteria analysis for this example.

	Performance Criteria	Pass or NYA
1	techniques appropriate to the task are selected and used by the individual	Pass used addition or subtraction
2	activities required by the techniques are performed to appropriate levels of precision and in the correct sequence	Pass added minutes first then hours
3	mathematical terms and mathematical aspects of everyday language are used and interpreted with precision (AT1 3b)	Pass used + and –
4	data are recorded in the required units	Pass used hours and minutes
5	data are recorded in the required format	Pass numbers
6	records are accurate and complete	Pass all correct

Fig 1.2

Exercise 1.3

Estimate your answer before calculating.

1 Add the following times:

(a) 1 hr 30 min + 2 hr 20 min
(b) 55 sec + 1 min 5 sec
(c) 13 hr 38 min + 5 hr 53 min
(d) 2 days 3 hr + 23 hr

2 Subtract the following times:

(a) 5 hr 50 min – 3 hr 20 min
(b) 3 days – 12 hr
(c) 1 min 15 sec – 45 sec
(d) 2 hr 40 min 30 sec – 1 hr 22 min 13 sec

2 Convert between different units of measurement using tables, graphs and scales

In the previous chapter it was mentioned that there are people who still use imperial units of measurement. Because both forms are still in use it is important that you can convert between the two. Conversion tables, graphs and scales will help you do this, but you should also know the more common measures in your head.

You will find it helpful to learn the following:

2.54 cm = 1 inch
1 metre = 39.37 inches
1 kilometre = 0.6214 miles
28.35 grams = 1 oz
1 kilogram = 2.205 lb
1 litre = 0.22 gallons

Conversion tables

Conversion tables such as those that appear in Table 2.1 are extremely useful for quick reference.

Table 2.1

Inches		*Millimtrs*	*Feet*		*Metres*	*Miles*		*Kilomtrs*	*Yards*		*Metres*
0.039	**1**	25.400	3.281	**1**	0.305	0.621	**1**	1.609	1.094	**1**	0.914
0.079	**2**	50.800	6.562	**2**	0.610	1.243	**2**	3.219	2.187	**2**	1.829
0.118	**3**	76.200	9.842	**3**	0.914	1.864	**3**	4.828	3.281	**3**	2.743
0.158	**4**	101.600	13.123	**4**	1.219	2.486	**4**	6.437	4.375	**4**	3.658
0.197	**5**	127.000	16.404	**5**	1.524	3.107	**5**	8.047	5.468	**5**	4.572
0.236	**6**	152.400	19.685	**6**	1.829	3.728	**6**	9.656	6.562	**6**	5.486
0.276	**7**	177.800	22.966	**7**	2.134	4.350	**7**	11.265	7.655	**7**	6.401
0.315	**8**	203.200	26.246	**8**	2.438	4.971	**8**	12.875	8.749	**8**	7.315
0.354	**9**	228.600	29.527	**9**	2.743	5.592	**9**	14.484	9.843	**9**	8.230

Ounces		*Grams*	*Pounds*		*Kilogrms*	*Gallons*		*Litres*	*Pints*		*Litres*
0.035	**1**	28.350	2.205	*1*	0.454	0.220	**1**	4.546	1.760	**1**	0.568
0.071	**2**	56.699	4.409	**2**	0.907	0.440	**2**	9.092	3.520	**2**	1.137
0.106	**3**	85.049	6.614	**3**	1.361	0.660	**3**	13.638	5.279	**3**	1.705
0.141	**4**	113.398	8.819	**4**	1.814	0.880	**4**	18.184	7.039	**4**	2.273
0.176	**5**	141.748	11.023	**5**	2.268	1.100	**5**	22.730	8.799	**5**	2.841
0.212	**6**	170.097	13.228	**6**	2.722	1.320	**6**	27.277	10.559	**6**	3.410
0.247	**7**	198.447	15.432	**7**	3.175	1.540	**7**	31.823	12.318	**7**	3.978
0.282	**8**	226.796	17.637	**8**	3.629	1.760	**8**	36.369	14.078	**8**	4.546
0.318	**9**	255.146	18.842	**9**	4.082	1.980	**9**	40.915	15.838	**9**	5.114

The numbers in bold in the centre of each table can indicate either measure.

EXAMPLE 1

(a) Convert 1 ounce into grams.

Ounces		*Grams*
0.035	↘ **1** →	28.350
0.071	**2**	56.699

Read from left to right.

Ounces → **1** → 28.350 g

(b) Convert 2 grams into ounces.

Ounces		*Grams*
0.035	**1** ↙	28.350
0.071	← **2**	56.699

Read from right to left.

Grams → **2** → 0.071 oz

Figure 2.1 gives the performance criteria analysis for this example.

	Performance Criteria	Pass or NYA
1	techniques appropriate to the task are selected and used by the individual	Pass used appropriate table
2	activities required by the techniques are performed to appropriate levels of precision and in the correct sequence	Pass a) read from left to right b) read from right to left
3	mathematical terms and mathematical aspects of everyday language are used and interpreted with precision (AT1 3b)	Pass used the verb "to convert"
4	data are recorded in the required units	Pass g and oz
5	data are recorded in the required format	Pass numbers
6	records are accurate and complete	Pass correct answers

Fig 2.1

Exercise 2.1

1 Convert the following:

(a) 8 feet into metres
(b) 4 kilograms into pounds
(c) 5 litres into gallons

2 A filing cabinet weighs 70 kilograms. How much is this in pounds?

3 A canister holds 52 litres of water. How much is this in gallons?

4 My pen is 4 inches long. How many millimetres is this?

5 The distance from Redcar to Guisborough is 9 miles. How far is this in kilometres?

Conversion formulae

Sometimes it is easier to work with conversion formulae tables (see Table 2.2).

Table 2.2

To convert	*Multiply by*	*To convert*	*Multiply by*
Inches to centimetres	2.54	Acres to hectares	0.4047
Centimetres to inches	0.3937	Hectares to acres	2.471
Feet to metres	0.3048	Gallons to litres	4.546
Metres to feet	3.281	Litres to gallons	0.22
Yards to metres	0.9144	Ounces to grams	28.35
Metres to yards	1.094	Grams to ounces	0.03527
Miles to kilometres	1.609	Pounds to kilograms	0.4536
Kilometres to miles	0.6214	Kilograms to pounds	2.205

EXAMPLE 2

Convert 100 litres to gallons.

From the right side of the table we read that to convert litres to gallons you multiply by 0.22.

$100 \times 0.22 = 22$

Answer = 22 gallons

Figure 2.2 gives the performance criteria analysis for this example.

	Performance Criteria	Pass or NYA
1	techniques appropriate to the task are selected and used by the individual	Pass used correct conversion formula
2	activities required by the techniques are performed to appropriate levels of precision and in the correct sequence	Pass used litres x by conversion
3	mathematical terms and mathematical aspects of everyday language are used and interpreted with precision (AT1 3b)	Pass used verb "to convert" Used multiplication
4	data are recorded in the required units	Pass gallons
5	data are recorded in the required format	Pass numbers
6	records are accurate and complete	Pass correct answer

Fig 2.2

Exercise 2.2

1 A petrol tank holds 10 gallons. How many litres will it hold?

2 If a journey is 48 miles, how many kilometres is it?

3 A farmer has 40 acres of land. What is this in hectares?

4 A block of margarine weighs 500 grams. How much does it weigh in ounces?

5 An elderly lady wishes to buy a washing machine 60 cm wide. She does not know if this will fit into her kitchen as she has measured the gap in inches. Convert 60 cm to inches.

Conversion Scales

Figures 2.3 to 2.7 show the conversion scales of some of the most common units of measurement.

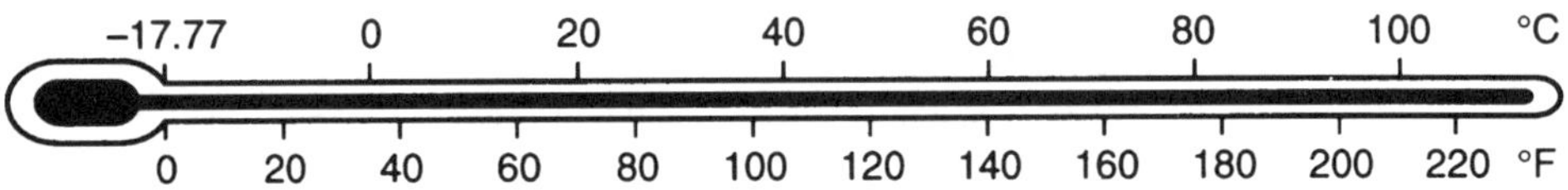

Fig 2.3 Conversion scale for °C to°F

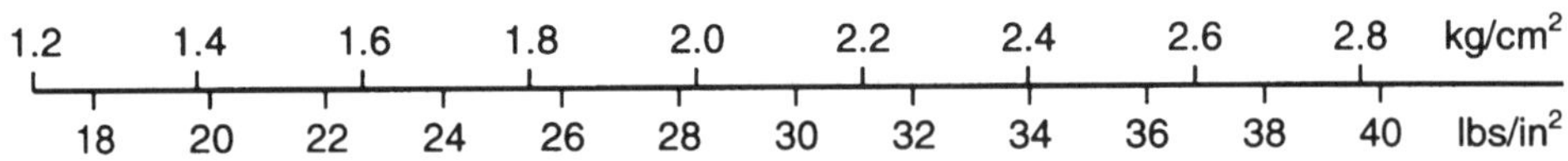

Fig 2.4 Conversion scale for kg/cm^2 to lbs/in^2

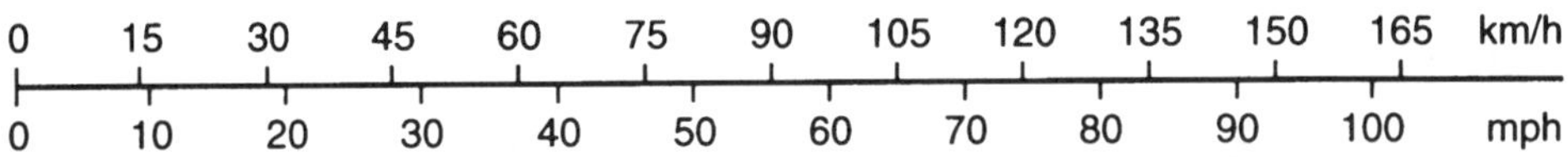

Fig 2.5 Conversion scale for km/h to mph

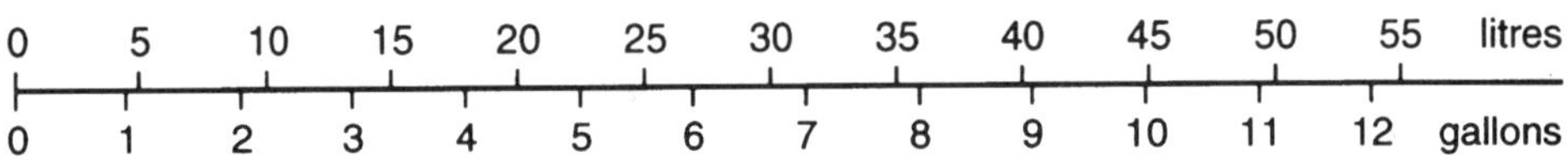

Fig 2.6 Conversion scale for litres to gallons

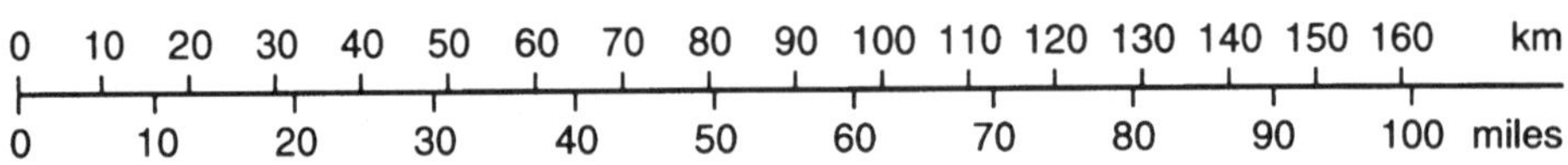

Fig 2.7 Conversion scale for km to miles

EXAMPLE 3

Convert 40 mph to km/h.

Read along the bottom line of Fig 2.5 to 40 mph.

The answer is 64 km/h.

Figure 2.8 gives the performance criteria analysis for this example.

	Performance Criteria	Pass or NYA
1	techniques appropriate to the task are selected and used by the individual	Pass used correct conversion scale
2	activities required by the techniques are performed to appropriate levels of precision and in the correct sequence	pass. Read along top line to 40 mph
3	mathematical terms and mathematical aspects of everyday language are used and interpreted with precision (AT1 3b)	Pass. Used verb "to convert"
4	data are recorded in the required units	Pass km/h
5	data are recorded in the required format	Pass numbers
6	records are accurate and complete	Pass All correct

Fig 2.8

Remember, scales are not as accurate as tables. Sometimes your answer will have to be your best educated guess.

Exercise 2.3

Using the conversion scales, answer the following questions.

1 A car travels on the motorway at 112 km/h. What is its speed in mph?

2 The pressure of each tyre on a car should be 24 lb/in^2. What is this in kg/cm^2?

3 An elderly woman has a body temperature of 95 °F. What is this in °C?

4 A tank holds 12 gallons of water. What is this in litres?

5 It is 70 miles from Leeds to Derby. What is this in km?

Conversion Graphs

Figures 2.9 to 2.12 show the conversion graphs for some common units of measurement.

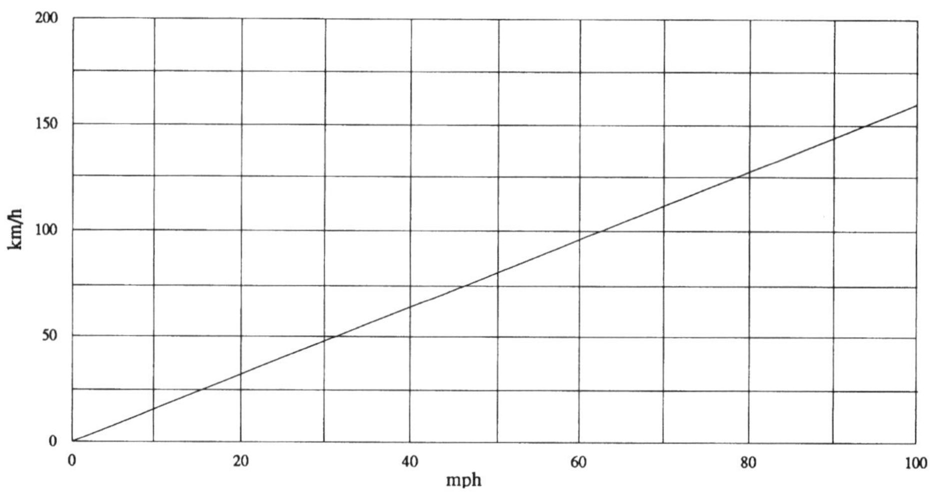

Fig 2.9 Conversion graph for mph to km/h

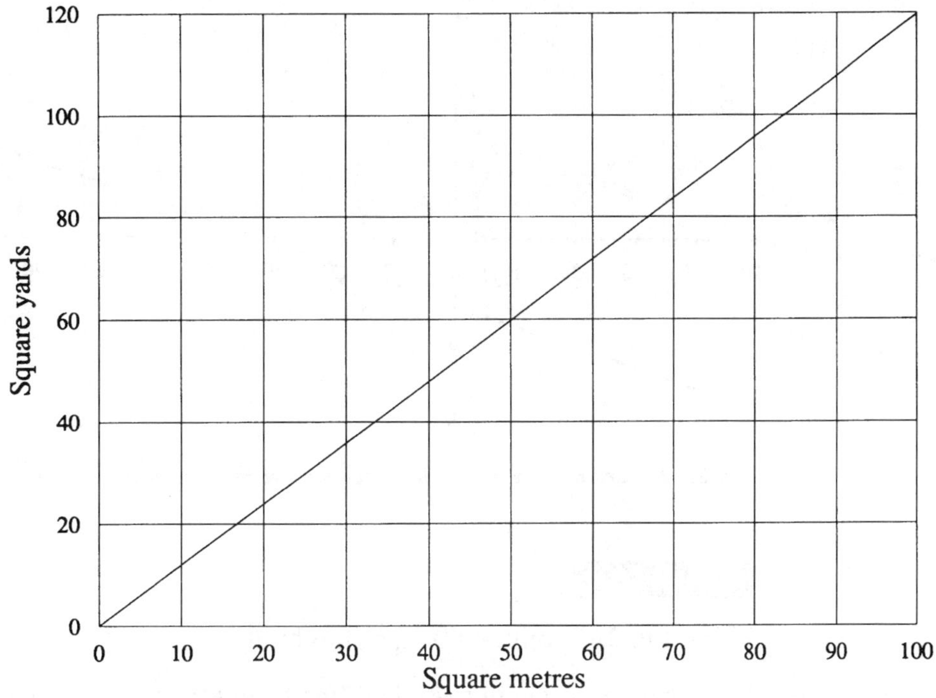

Fig 2.10 Conversion graph for square metres to square yards

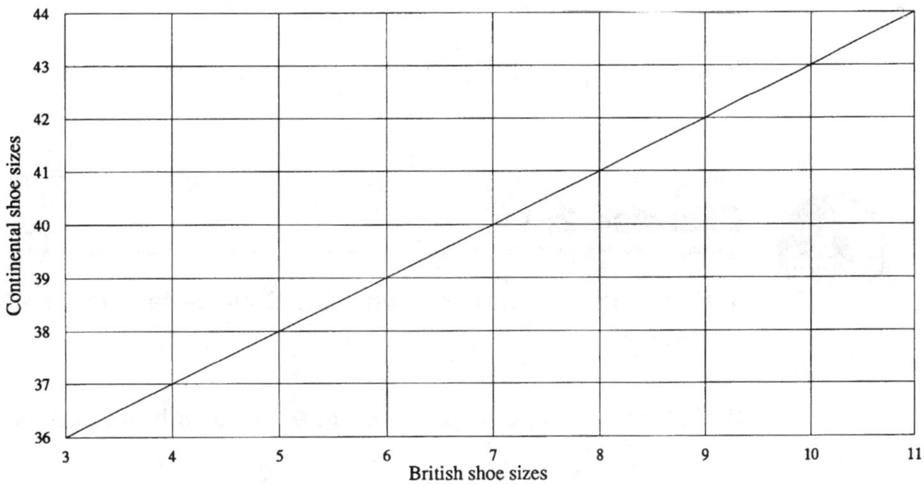

Fig 2.11 Conversion graph for British and Continental shoe sizes

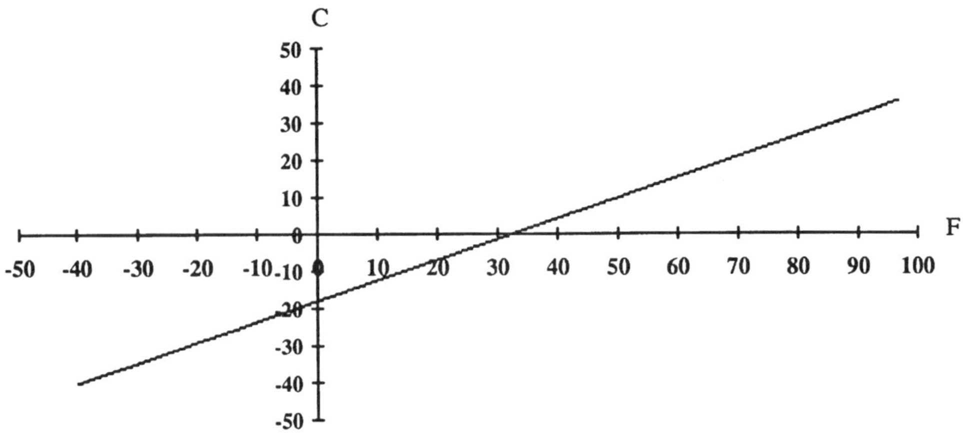

Fig 2.12 Conversion graph for Fahrenheit to Centigrade

EXAMPLE 4

Using Fig 2.9, convert 60 mph into km/h.

Use a ruler to draw in a vertical line from 60 on the x axis to the graph line. Draw a horizontal line from the graph line to the y axis. Read the value off the y axis.

Answer is 97 km/h.

Figure 2.13 gives the performance criteria analysis for this example.

Exercise 2.4

1 Using Fig 2.11, work out what the Continental size for a British shoe size 6 is.

2 Using Fig 2.10, convert 45 square metres into square yards.

3 Using Fig 12.12, express 50 °F in °C.

4 A salesman has a choice between two company cars. Car A accelerates from 0 to 60 mph in 9 seconds. Car B accelerates from 0 to 100 km/h in 9 seconds. Which car is faster?

	Performance Criteria	Pass or NYA
1	techniques appropriate to the task are selected and used by the individual	Pass used correct conversion graph
2	activities required by the techniques are performed to appropriate levels of precision and in the correct sequence	Pass used ruler to draw line on graph to read answer
3	mathematical terms and mathematical aspects of everyday language are used and interpreted with precision (AT1 3b)	Pass was able to carry out Pc 1 and 2
4	data are recorded in the required units	Pass Km/h
5	data are recorded in the required format	Pass numbers
6	records are accurate and complete	Pass correct answer

Fig 2.13

5 A children's home requires new carpet in the day room. The room measures 80 yds^2. The showroom has two carpets that are ideal. Carpet A is £5 per m^2 and carpet B is a remnant of 70 m^2 costing £340. Which is the cheaper carpet?

6 A friend from the USA boasted that he sat through an American Football game when the temperature was – 14°F. During the Winter Olympics you sat through temperatures reaching – 20°C. Which is colder?

3 Conduct a survey on an issue of the individual's choice

A survey is a way of collecting statistical data. It can be something very simple such as a show of hands or a more complicated method such as a written questionnaire.

There are five common survey methods:

1 Show of hands
2 Face to face interview
3 Telephone interview
4 Questionnaire
5 Postal questionnaire.

The most appropriate method for your survey will depend upon the complexity of the issue you choose.

Show of hands

Show of hands is a very easy-to-use method with instant results. The surveyor asks a question and each respondent raises his or her hand. The raised hands are then counted.

■ Disadvantages

1 Its scope is limited.
2 It is sometimes viewed as being childish.
3 Many people are shy about personal or confidential information.

EXAMPLE 1

A survey of employment patterns was conducted among 20 students. The question was asked, 'If you have a regular part-time job, raise your hand'. The raised hands were counted. 15 people out of 20 have a regular part-time job.

Figure 3.1 shows the performance criteria analysis of this example.

	Performance Criteria	Pass or NYA
1	techniques appropriate to the task are selected and used by the individual	Pass. Show of hands method
2	activities required by the techniques are performed to appropriate levels of precision and in the correct sequence	Pass. Asked question; recorded result
3	mathematical terms and mathematical aspects of everyday language are used and interpreted with precision (AT1 3b)	Pass. Used show of hands as a survey method
4	data are recorded in the required units	Pass people
5	data are recorded in the required format	Pass numbers
6	records are accurate and complete	Pass. Contains all information

Fig 3.1

Face-to-face interview

Many market research companies use face-to-face interviews to survey the public. The interviewers are trained and fairly well paid, which makes this survey method costly.

Advantages

1 The interviewer can show examples of the product, advertisement, article being surveyed.

2. The interviewer can ask more detailed questions or explain difficult questions.
3. The interviewer can watch for exaggerations or obvious lies (e.g. a man wearing an overcoat in the office and stating that the temperature is comfortable).
4. More time can be spent as people do not wish to be impolite and cut short the interview.
5. It is possible to obtain more interviews as many people cannot say no when confronted face-to-face.

Disadvantages

1. Interviewers can introduce bias into the survey, by the way they ask the questions or the way they record the answers.
2. The method is also time consuming as each interview can take up to 10 minutes.

EXAMPLE 2

A survey was conducted into the taste preference between two similar fizzy drinks. People were asked the following questions:

- 'Which drink do you prefer, A or B?'
- 'Do you regularly buy the drink you have chosen?'
- 'Would you buy it if you were given a money-off voucher?'

The survey resulted in the following:

- 55% of the people surveyed preferred Drink A, 40% preferred Drink B, and 5% did not like either drink.
- 30% of the people surveyed said they bought their preferred drink regularly.
- 80% of the people surveyed accepted a money-off voucher.

Figure 3.2 gives the performance criteria analysis for this example.

	Performance Criteria	Pass or NYA
1	techniques appropriate to the task are selected and used by the individual	Pass. Used face-to-face interview
2	activities required by the techniques are performed to appropriate levels of precision and in the correct sequence	Pass. Asked questions and recorded answers
3	mathematical terms and mathematical aspects of everyday language are used and interpreted with precision (AT1 3b)	Pass. Used face-to-face interview as a survey method
4	data are recorded in the required units	Pass People surveyed
5	data are recorded in the required format	Pass. Wrote answers down and used percentages
6	records are accurate and complete	Pass. Recorded answers of all surveyed

Fig 3.2

Telephone interview

The telephone interview is less costly than the face-to-face interview. However, it is easier for someone to decline an interview by replacing the receiver. Other disadvantages are that the interview is limited by time (because of the cost of the call) and limited to people with telephones.

EXAMPLE 3

A survey was conducted into the swimming habits of households. The surveyor began by choosing 20 telephone numbers at random. Each telephone call was started with the surveyor saying, 'I am conducting a survey into the swimming habits of households.' People were then asked the following questions:

- 'Does any member of your family regularly participate in swimming?'
- 'How often do the members of your family participate in swimming?'
- 'Does any member of your family receive swimming lessons?'

- 'Does your family prefer to attend a regular swimming pool or one with a wave machine and water chute?'

The survey resulted in the following:

- 35% of the people surveyed have a family member who regularly participates in swimming.
- The most common answer to how often they participated in swimming was once a week.
- 20% of the families surveyed have a family member who receives swimming lessons.
- 60% of the families surveyed preferred to attend a pool with a wave machine and water chute.

Figure 3.3 gives the performance criteria analysis for this example.

	Performance Criteria	Pass or NYA
1	techniques appropriate to the task are selected and used by the individual	Pass. Used Telephone Interview
2	activities required by the techniques are performed to appropriate levels of precision and in the correct sequence	Pass. Chose sample frame first, then surveyed people
3	mathematical terms and mathematical aspects of everyday language are used and interpreted with precision (AT1 3b)	Pass. Used telephone Interview as a Survey method
4	data are recorded in the required units	Pass. Used people surveyed
5	data are recorded in the required format	Pass. wrote answers down and used %
6	records are accurate and complete	Pass. Recorded answers of all surveyed.

Fig 3.3

Questionnaire

Most questionnaires are given out in stores or magazines.

■ Advantages

1 A person can answer the questionnaire in his or her own time.
2 Personal questions can be answered in confidence (no one need know who completed the questionnaire).

■ Disadvantages

1 There is no one to explain difficult questions.
2 The response rate is very poor.
3 The questionnaire has to be well designed to make analysis possible.
4 The questionnaire may be completed by a group of people thus making the information invalid.

Postal questionnaire

A postal questionnaire has the same advantages and disadvantages as a questionnaire. The response rate, however, is even poorer (about 2–10%).

EXAMPLE 4

Car owners received the questionnaire shown in Fig 3.4 through the post three weeks after having their car serviced.

Figure 3.5 shows the extent to which the questionnaire meets the performance criteria.

Do you still own the car that was recently serviced by us?	Yes ☐
	No ☐

Are we the dealership that you purchased the car from originally?	Yes ☐
	No ☐

Please indicate by ticking the appropriate box how much you agree or disagree with the following statements.

	Agree strongly	Agree slightly	Neither agree nor disagree	Disagree slightly	Disagree strongly
I was attended to within a reasonable length of time	☐	☐	☐	☐	☐
The staff were courteous and helpful at all times	☐	☐	☐	☐	☐
The mechanic understood my requirements	☐	☐	☐	☐	☐
My car was left in safe and competent hands	☐	☐	☐	☐	☐
All my instructions were carried out satisfactorily	☐	☐	☐	☐	☐
The work was carried out to a high standard	☐	☐	☐	☐	☐
It is likely that I shall return for my next service	☐	☐	☐	☐	☐

Fig 3.4 Example of a postal questionnaire

	Performance Criteria	Pass or NYA
1	techniques appropriate to the task are selected and used by the individual	Pass. Postal Questionnaire
2	activities required by the techniques are performed to appropriate levels of precision and in the correct sequence	Pass. Posted questionnaire to appropriate sample
3	mathematical terms and mathematical aspects of everyday language are used and interpreted with precision (AT1 3b)	Pass. Used postal questionnaire as a survey method
4	data are recorded in the required units	NYA. Data not yet analysed
5	data are recorded in the required format	NYA. Data not yet recorded
6	records are accurate and complete	NYA. Not yet complete

Fig 3.5

4 Design and use an observation sheet to collect data

Observation is a valid and common method of data collection. Some data can only be collected by observation: for example, how many men and how many women attend a certain theatre performance. The total number of people in attendance can be taken from the tickets sold. However, the breakdown of men and women can only be carried out by observation (unless customers buy male and female tickets). Much observation can be performed by computer now; although the trained observer is still required in many industries.

Observation is a record of what happened; for example, how long it took a team of five men to build a brick wall 12 feet high and 40 feet long. It does not record whether the men thought the task to be strenuous or difficult, but it may record how many bricks and how much mortar they used.

Observation can be used to establish whether working patterns or communication methods in an organisation are adequate. For this kind of observation the people being observed must remain unaware of the observer. This is therefore a disadvantage as the observer must be trained to remain inconspicuous. Another disadvantage of observation is that it tells us nothing about people's attitudes (this can however be established in a questionnaire). Observation also gives us no indication of past or future data.

A tally chart

A tally chart is the ideal method for recording observed data.

EXAMPLE 1

A care assistant observes the stock of cereal in a children's home. Her tally chart appears in Fig 4.1.

Cereal	Tally	Frequency
Cornflakes	~~IIII~~ III	8
Shreddies	~~IIII~~ ~~IIII~~	10
Rice Crispies	III	3
Sugar Puffs	~~IIII~~ ~~IIII~~ II	12
Muesli	~~IIII~~ IIII	9
Ready Break	~~IIII~~ ~~IIII~~ ~~IIII~~	15

Fig 4.1 A tally chart

A vertical line is given for each packet of cereal. Four lines are grouped with a fifth horizontal line. This is the tally. Frequency is the tally converted into whole numbers.

Figure 4.2 shows the performance criteria analysis for this example.

	Performance Criteria	Pass or NYA
1	techniques appropriate to the task are selected and used by the individual	Pass used a tally chart
2	activities required by the techniques are performed to appropriate levels of precision and in the correct sequence	Pass tally chart formed correctly
3	mathematical terms and mathematical aspects of everyday language are used and interpreted with precision (AT1 3b)	Pass used the verb "to observe" correctly
4	data are recorded in the required units	Pass. Frequency
5	data are recorded in the required format	Pass numbers and tally
6	records are accurate and complete	Pass Correct answer

Fig 4.2

A two-way table

A two-way table is another method of recording observed data.

EXAMPLE 2

The output of five assembly workers was observed over a week and recorded on an observation sheet (see Fig 4.3).

Worker \ Day	Mon	Tue	Wed	Thur	Fri
M Jones	157	155	157	150	160
K Fitz	170	169	166	170	168
J Clark	165	164	164	164	164
S Oliver	147	148	148	148	147
D Simpson	150	150	151	150	149

Fig 4.3 A two-way table

The performance criteria analysis for this example is shown in Fig 4.4.

	Performance Criteria	Pass or NYA
1	techniques appropriate to the task are selected and used by the individual	Pass. Used a two-way table
2	activities required by the techniques are performed to appropriate levels of precision and in the correct sequence	Pass. Two-way table formed correctly
3	mathematical terms and mathematical aspects of everyday language are used and interpreted with precision (AT1 3b)	Pass "to observe" correctly
4	data are recorded in the required units	Pass. recorded as numbers
5	data are recorded in the required format	Pass. recorded on table
6	records are accurate and complete	Pass Complete table

Fig 4.4

Exercise 4.1

1 A car park attendant observed the length of stay of 30 cars to the nearest 15 minutes. Produce a tally chart and frequency distribution for this information.

15	30	60	15	75	45	75	120	30	90
30	120	75	90	135	45	75	105	45	60
60	75	120	45	75	60	90	75	60	120

2 Design a two-way table for a supervisor of five labourers. The supervisor wishes to observe the time it takes each labourer to accomplish the following activities:

- Lay a course of bricks 10 m long
- Mix a metre of concrete
- Dig two fence post holes
- Erect 5 m of scaffolding.

3 Design and use a tally chart to observe the types and number of each type of fire extinguisher in a building.

4 Design and use a two-way table to observe the prices of five different products in three competing stores.

5 Design and use a questionnaire to survey opinion

The advantages and disadvantages of a questionnaire have been discussed in Chapter 3. This chapter will concentrate on the design and use of the questionnaire.

Design of questionnaire

It is important to remember when designing a questionnaire that the questionnaires have to be analysed. If 50 questionnaires are returned, there has to be a system of analysing them. When you design a questionnaire, you should have an idea of the kind of answers you are likely to receive.

Here are some rules to remember when designing a questionnaire.

1. Keep the questionnaire short (one or two pages).
2. Start with very simple questions.
3. Give as many codified response questions as possible (questions with a range of answers to choose from).
4. Keep open-ended or complicated questions to a minimum. Better still have none at all.
5. Keep complicated or delicate questions to the end (name, address, telephone number, salary, etc. are all delicate questions).

EXAMPLE 1

(a) Start with very simple questions (see Fig 5.1).

Do you still own the car that was recently serviced by us?	Yes	☐
	No	☐

Fig 5.1

(b) Give as many codified response questions as possible (see Fig 5.2).

Please indicate by ticking the appropriate box how much you agree or disagree with the following statements.

	Agree strongly	Agree slightly	Neither agree nor disagree	Disagree slightly	Disagree strongly
I was attended to within a reasonable length of time	☐	☐	☐	☐	☐
The staff were courteous and helpful at all times	☐	☐	☐	☐	☐
The mechanic understood my requirements	☐	☐	☐	☐	☐
My car was left in safe and competent hands	☐	☐	☐	☐	☐
All my instructions were carried out satisfactorily	☐	☐	☐	☐	☐
The work was carried out to a high standard	☐	☐	☐	☐	☐
It is likely that I shall return for my next service	☐	☐	☐	☐	☐

Fig 5.2

(c) Keep open-ended or complicated questions to a minimum (see Fig 5.3).

Which aspect of service would you like to see improved?

____

Fig 5.3

(d) Keep complicated or delicate questions to the end (see Fig 5.4).

What are your views on euthanasia?

____

Do you know why the House of Lords voted against euthanasia?

____

Fig 5.4

Figure 5.5 gives the performance criteria analysis for this example.

	Performance Criteria	Pass or NYA
1	techniques appropriate to the task are selected and used by the individual	Pass. Correct type of questions for questionnaire
2	activities required by the techniques are performed to appropriate levels of precision and in the correct sequence	Pass. Questions in the correct order
3	mathematical terms and mathematical aspects of everyday language are used and interpreted with precision (AT1 3b)	NYA. Questions are in the design stage only. Not yet used.
4	data are recorded in the required units	
5	data are recorded in the required format	
6	records are accurate and complete	

Fig 5.5

Use of questionnaire

Who will we give the questionnaire to? This is a very good question. It is of no use to give a questionnaire intended for office workers to shop assistants. It is also very time consuming and costly to give a questionnaire to everyone in a large organisation.

■ Population

Population is a term used by statisticians. It refers to any group of items from which a sample may be taken. Some examples of population are:

- The employees of a certain company
- The subscribers to a certain magazine
- Boxes of muesli on a production line
- Dresses made in a factory.

Sampling frame

The sampling frame is a complete list or complete identification of the population. Some examples of sampling frame are:

- The names and departments of employees in a certain company. (This information can be obtained from personnel records.)
- The names and addresses of subscribers to a certain magazine. (This information can be obtained from customer records.)
- Every box of muesli coming off a production line in any one day.
- Every dress made in a factory in any one day.

Sampling techniques

Sampling techniques is the final stage used to identify exactly to whom the questionnaire should be given. There are five common sampling techniques:

- Simple random sampling
- Stratified sampling
- Systematic sampling
- Multi-stage sampling
- Quota sampling.

Simple random sample

This technique is often know as the 'lottery method'. It is so named because it follows this procedure:

1. All the items from the sample frame are given a number.
2. The numbers are placed in a hat or box.
3. A proportion of numbers is chosen at random.
4. The chosen numbers indicate who should be given the questionnaire to complete.

This procedure is acceptable if you are only dealing with a small sample frame. A larger sample frame will require the use of a computer generated random table.

Stratified sample

The word stratified means to form into layers or status groups. A stratified sample is therefore one that divides the sample frame into groups. Some examples of stratified groups are:

- Full-time and part-time workers
- Male and female
- Under 18 years and over 18 years
- Cars with four doors and cars with less than four doors
- Biscuits with cream, biscuits with chocolate and biscuits with cream *and* chocolate
- Accounts department, personnel department, marketing department, administration department, etc.

This technique can be divided into three discrete steps.

1. The sample frame is divided into the required groups.
2. The proportion of each group to the whole is calculated (e.g. 120 male workers and 80 female workers in an organisation will be 60% male and 40% female).
3. A simple random sample is taken from each group. (This is also calculated by proportion, e.g. of the 60% male workers 6 may be questioned and of the 40% female workers 4 may be questioned.)

Systematic sample

This technique is one of the easiest and least expensive of all the techniques. From the sample frame every *n*th item is chosen. Some examples of systematic sampling are:

- If the sample frame is a list of 100 workers and 10% are to be questioned every 10th worker is chosen.
- If the sample frame is the telephone directory, open any page and choose every 20th person.
- From a production line every 50th toy is checked for faults. (The disadvantage of using this technique for a production line is that the machinery could be faulty at regular intervals.)

Multi-stage sample

Multi-stage sampling is mainly used when the sample frame is very large or geographically spread out. For example, if the Department of Transport wished to survey drivers' views, the sample frame would be every adult with a current driving licence. To random sample this population could mean interviewers going to many parts of the country. This would be a very costly and time consuming exercise. Multi-stage sampling will reduce the costs and time required.

The procedure the technique follows is:

1 The country is divided into a few large regions. (Independent Television regions are a popular choice of market research companies.)
2 Two or three of these large regions are chosen at random.
3 The two or three chosen regions are subdivided into smaller areas. (Parliamentary constituencies or local authority regions are a common choice.)
4 A few of these smaller areas are chosen at random.
5 From a map of the few chosen smaller areas streets are chosen at random.
6 House numbers from each chosen street are random sampled until a list of individuals is obtained.

Figure 5.6 illustrates the procedure diagrammatically.

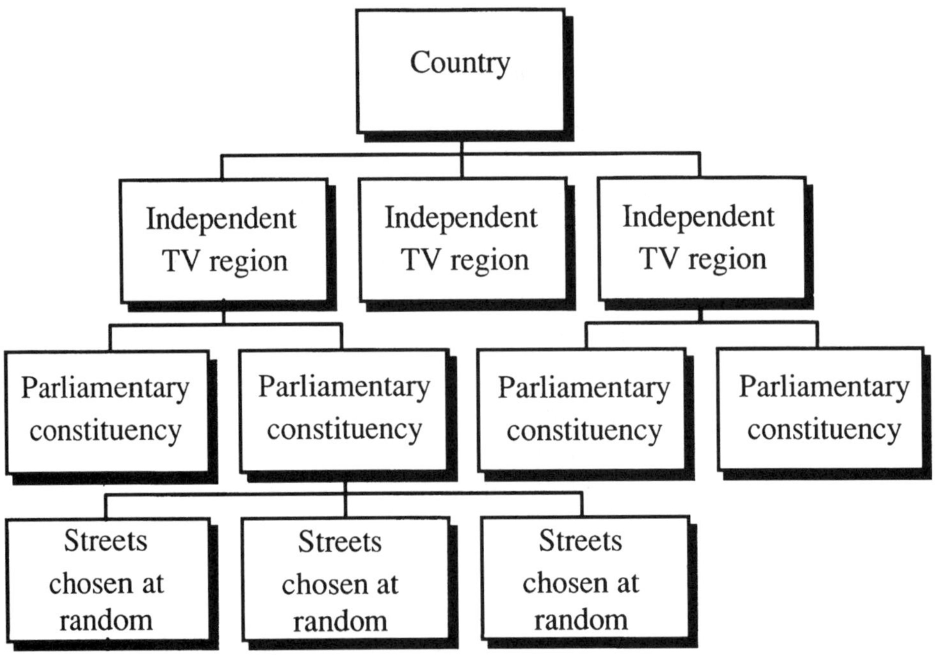

Fig 5.6 Multi-stage sampling

Quota sample

This technique will not give a great deal of accuracy. However, it is popular with market research companies as it is easy to implement. The 'quota' is a set number of a particular group from the sample frame. Some examples of quota samples are:

- 30 women aged 18 to 30.
- 20 American males.
- 30 unemployed men and 20 unemployed women aged 16 to 20.
- 100 white-collar males aged 30 to 65.
- 20 grandmothers.

6 Organise data into groups and classifications

You are required to organise some collected data into groups and classifications. It is more likely to be the data you collected from your questionnaire rather than the data collected for the survey or from observation. The purpose of organising data is to make any analysis of the questionnaire or drawing of conclusions easier.

First the difference between groups and classifications needs to be established. *Classification* is the organisation of objects into classes according to their similarities or differences. Each class must be independent of the next class. An object cannot be placed into more than one class. *Groups* are those items on your questionnaire that can be bound together in any way. An object can be placed into one or more groups.

Data can be organised into a tally chart, two-way table or scatter graph.

Tally charts and two-way tables

The design of a tally chart and a two-way table were covered in Chapter 4. Here is an example of how to display results from a questionnaire.

EXAMPLE 1

Ten nurses were questioned about hospital administration systems. The nurses were asked their views on the efficiency of the administration systems. The results were displayed on a two-way table (see Fig 6.1).

Administration system \ Nurse	1	2	3	4	5	6	7	8	9	10
Patients records	VG	G	VG	B	VG	G	A	G	VG	A
Medication stock	VG	G	G	B	VG	G	A	VG	VG	A
Equipment stock	VG	G	A	B	B	G	G	VG	VG	A
Doctors roster	G	G	A	G	A	G	G	VG	G	B
Nurses roster	G	A	A	VB	VG	G	A	B	VG	A

VG = Very Good ; G = Good ; A = Average ; B = Bad ; VB = Very Bad

Fig 6.1 Example of a two-way table

Figure 6.2 shows the performance criteria analysis.

	Performance Criteria	Pass or NYA
1	techniques appropriate to the task are selected and used by the individual	Pass. Two-way table used to display results
2	activities required by the techniques are performed to appropriate levels of precision and in the correct sequence	Pass. Two-way table designed correctly
3	mathematical terms and mathematical aspects of everyday language are used and interpreted with precision (AT1 3b)	Pass. Good interpretation of requirements
4	data are recorded in the required units	Pass. Used abreviations
5	data are recorded in the required format	Pass letters
6	records are accurate and complete	Pass Complete

Fig 6.2

A scatter graph

A scatter graph is an XY graph where the X axis is the independent variable and the Y axis the dependent variable. Each point is plotted on the graph, but the points are not joined up.

A scatter graph is used to show a *correlation* between data. Correlation means a connection or a relationship between data.

EXAMPLE 2

A factory worker was observed over a week. The chart in Fig 6.3 shows the number of items made during each hour. The information was then displayed as a scatter graph (see Fig 6.4).

Day \ Hours	10	11	12	2	3	4	5
Monday	150	148	147	140	130	120	100
Tuesday	148	147	147	140	135	130	128
Wednesday	155	150	140	135	125	120	120
Thursday	145	145	140	135	130	130	110
Friday	130	128	127	120	110	108	100

Fig 6.3 Chart for number of items made

The scatter graph in Fig 6.4 shows that the worker was able to make more items at the beginning of the day and less at the end of the day. Could this possibly be because of tiredness in the afternoon?

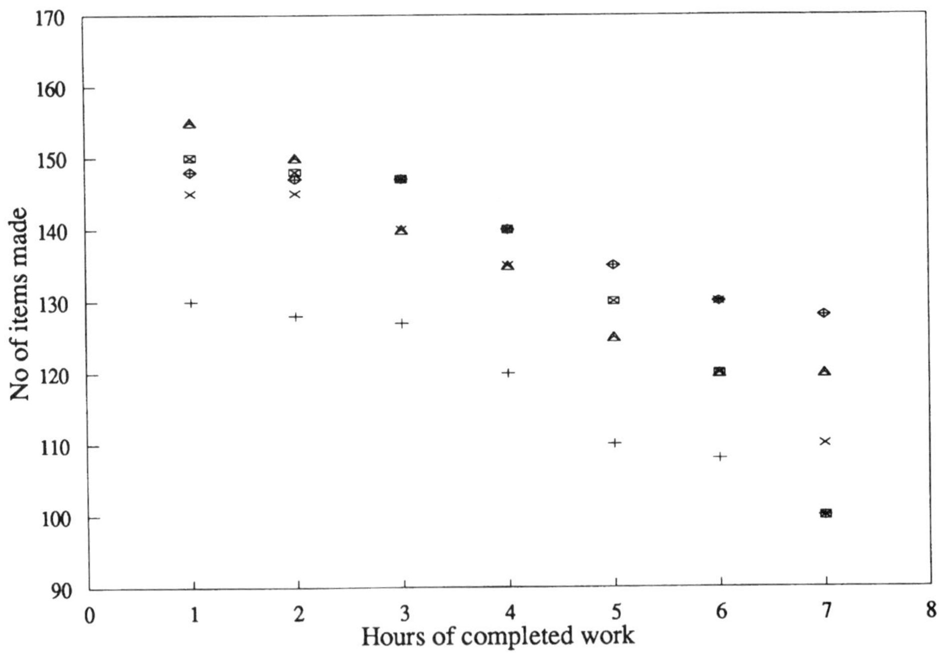

Fig 6.4 Scatter graph showing number of items made during each hour

Figure 6.5 shows the performance criteria analysis for this example.

	Performance Criteria	Pass or NYA
1	techniques appropriate to the task are selected and used by the individual	Pass. Used Two-way table and scatter graph
2	activities required by the techniques are performed to appropriate levels of precision and in the correct sequence	Pass. Correct layout for table and graph
3	mathematical terms and mathematical aspects of everyday language are used and interpreted with precision (AT1 3b)	Pass. good interpretation of requirements
4	data are recorded in the required units	Pass. number of items
5	data are recorded in the required format	Pass. numbers and stars
6	records are accurate and complete	Pass Complete

Fig 6.5

Scatter graphs can be any one of three different types: negative correlation, positive correlation and no correlation. The scatter graph in Example 2 is a negative correlation. The more time the worker spends at work, the less work he does.

Figure 6.6 shows the three different types of scatter diagrams.

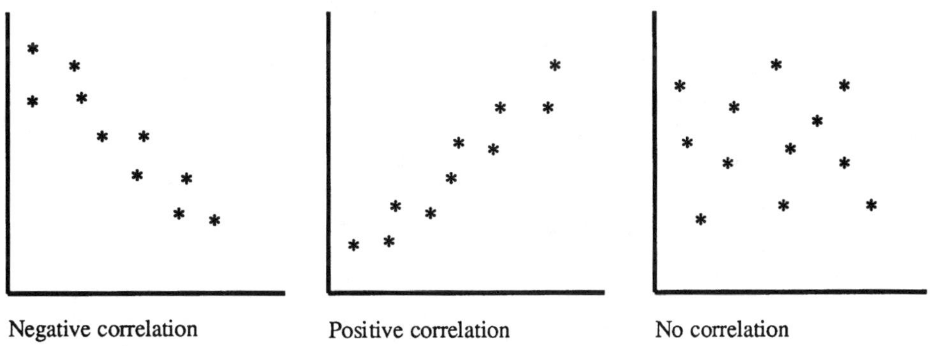

Fig 6.6 The three types of scatter graph

You will find more information on scatter graphs in Chapter 20.

Part 2
REPRESENT AND TACKLE PROBLEMS

Many problems in industry, business and public organisations can be solved using mathematical techniques: for example, is a machine still working effectively? This section introduces you to number rules, shape and space, and algebra as a way of solving problems.

7 Solve whole number problems involving addition and subtraction

After counting, addition and subtraction were the first mathematical techniques you learned as a young child. You will continue to use these techniques throughout your working life. It is therefore important that you can add and subtract both single and double digit numbers mentally. For example, if you have difficulty adding 10 + 35 + 44 mentally, you should start practising today. You should also be able to add slightly more difficult numbers without a calculator, e.g. using pencil-and-paper methods.

In the workplace it is not always possible to put your hands on a calculator immediately. It is therefore important that you can carry out calculations on paper or in your head. Even when you do have access to a calculator, it is still important to be able to add and subtract mentally to check that your answer is correct (you may press the wrong key, for instance).

Exercise 7.1

1 Calculate the following mentally:

(a) 5 + 7 + 9 =
(b) 14 + 3 + 7 =
(c) 9 + 21 + 2 =
(d) 4 + 7 + 12 + 8 + 16 + 3 =
(e) 6 + 18 + 24 + 35 + 7 =
(f) 25 – 12 =
(g) 40 – 14 =
(h) 29 – 17 =
(i) 49 – 15 =
(j) 14 – 6 =
(k) 17 – 9 =
(l) 35 – 23 =
(m) 65 – 29 =
(n) 24 – 15 =
(o) 16 – 7 =

2 A manufacturing company buys three items of office furniture: a desk costing £422, a chair costing £216 and a filing cabinet costing £462. How much did it spend altogether?

3 A nursing home buys new linen to replace items that have become old and worn out: 144 sheets, 96 duvet covers, 72 pillow cases, 24 table cloths and 12 teatowels. How many items were bought?

4 The receipts for a car service centre in one day's trading are shown in Table 7.1.

Table 7.1

Name	£	*Method of payment*
Mrs Norris	189.60	Cash
Mr Dixon	220.89	Cheque
The Ice-cream Centre	380.00	Account
Mr Spencer	55.00	Credit card
Miss Thompson	20.00	Cash
Mrs Evans	527.31	Credit card
Mr Tate	156.80	Cheque
Fox & Co.	150.00	Account
Mr Shaw	82.67	Cash
Mrs Grant	189.60	Credit card
Miss Holland	95.48	Cash
The Garden Centre	135.00	Account
Mrs Johnson	160.25	Cheque
Mr Stewart	672.87	Credit card

What were the following totals:

(a) Cash receipts
(b) Cheque receipts
(c) Credit card receipts
(d) Account receipts
(e) Total receipts?

5 The members of a fitness club held a sponsored step. Add up the number of steps achieved by each participant to get the total number for the event.

Pat Cummins	1360
John Hargreeves	1580
Amy Davies	1230
Jessica Long	1445
James Dawson	1400
Sam Wright	892

6 A manufacturer of kites employs five machinists to sew fabric stunt kites. From the weekly work sheet shown in Table 7.2 find:

(a) the machinist making the greatest number of kites
(b) the total number of kites made.

Table 7.2

	Mon	*Tues*	*Wed*	*Thur*	*Fri*
Machinist A	65	67	66	68	72
Machinist B	73	72	70	55	ill
Machinist C	60	69	70	75	65
Machinist D	71	71	71	71	71
Machinist E	58	58	59	57	60

7 During one day of trading an electrical superstore had the opening and closing stock levels shown in Table 7.3 (page 50).

Table 7.3

Item	*Opening stock*	*Closing stock*
Portable cassette player	49	37
Portable CD player	28	9
Mini hi-fi system	15	7
Midi hi-fi system	12	3
Car radio	13	9
Television	24	17
Video player	32	24
Satellite system	11	8
Washing machine	21	15
Tumble dryer	17	3
Refrigerator	18	12
Freezer	11	7
Vacuum cleaner	36	19

What quantity of each item was sold that day?

8 An upholsterer had 500 metres of brown fabric. He used 27 m to cover a three piece suite, 4.5 m to cover four dining room chairs, 8.5 m to cover a fireside chair and 16.25 m to cover a set of caravan cushions. How much fabric was remaining?

8 Solve problems involving multiplication and division

Along with addition and subtraction, you should be able to calculate multiplication and division mentally. This involves having instant recall of the 1 to 12 times tables and a working method of calculating higher tables. If you cannot recall all the 1 to 12 times tables then start practising now with the help of the tables on page 53.

Exercise 8.1

1 Mentally calculate the following:

(a) $6 \times 9 =$
(b) $3 \times 8 =$
(c) $4 \times 7 =$
(d) $9 \times 11 =$
(e) $12 \times 7 =$

(f) $8 \times 6 =$
(g) $9 \times 7 =$
(h) $4 \times 9 =$
(i) $7 \times 6 =$
(j) $9 \times 8 =$

(k) $36 \div 9 =$
(l) $24 \div 6 =$
(m) $21 \div 7 =$
(n) $220 \div 20 =$
(o) $49 \div 7 =$

2 The stock from The Bed Warehouse was damaged by floods, and had to be destroyed. From the record of items, quantity and selling price shown in Table 8.1, calculate how much the shop claimed from the insurers.

Table 8.1

Item	*Quantity*	*Selling price (£)*
2′ 6″ Divan	12	82.50
2′ 6″ Bunkbed	9	150.79
3′ Divan	35	109.20
3′ Bunkbed	27	160.85
Double divan	47	205.39
Double four poster	2	419.27
King size divan	19	375.80

3 A plumber needs to lay 27 m of 22 mm thick copper pipe and 132 m of 15 mm thick copper pipe for a central heating system. A length of pipe is 3 m long and costs £5.29 for 22 mm and £2.79 for 15 mm.

Calculate:

(a) how many lengths of 22 mm pipe are required;
(b) how many lengths of 15 mm pipe are required;
(c) the total cost.

4 A builder wishes to build a wall 6.16 m long by 1.04 m high. If a brick measures 22 cm by 6.5 cm, calculate the following.

(a) How many bricks are required?
(b) If a brick costs 15p, what is the cost of the bricks?

5 A theatre seats 1500 people. There are 250 box seats at £18.75 each, 450 stall seats at £16.25 and the remainder in the balcony at £13.50 each. If the theatre has eight shows in a week with all performances sold out, what would be the total revenue for that week?

1 x table

1	x	1	=	1
2	x	1	=	2
3	x	1	=	3
4	x	1	=	4
5	x	1	=	5
6	x	1	=	6
7	x	1	=	7
8	x	1	=	8
9	x	1	=	9
10	x	1	=	10
11	x	1	=	11
12	x	1	=	12

2 x table

1	x	2	=	2
2	x	2	=	4
3	x	2	=	6
4	x	2	=	8
5	x	2	=	10
6	x	2	=	12
7	x	2	=	14
8	x	2	=	16
9	x	2	=	18
10	x	2	=	20
11	x	2	=	22
12	x	2	=	24

3 x table

1	x	3	=	3
2	x	3	=	6
3	x	3	=	9
4	x	3	=	12
5	x	3	=	15
6	x	3	=	18
7	x	3	=	21
8	x	3	=	24
9	x	3	=	27
10	x	3	=	30
11	x	3	=	33
12	x	3	=	36

4 x table

1	x	4	=	4
2	x	4	=	8
3	x	4	=	12
4	x	4	=	16
5	x	4	=	20
6	x	4	=	24
7	x	4	=	28
8	x	4	=	32
9	x	4	=	36
10	x	4	=	40
11	x	4	=	44
12	x	4	=	48

5 x table

1	x	5	=	5
2	x	5	=	10
3	x	5	=	15
4	x	5	=	20
5	x	5	=	25
6	x	5	=	30
7	x	5	=	35
8	x	5	=	40
9	x	5	=	45
10	x	5	=	50
11	x	5	=	55
12	x	5	=	60

6 x table

1	x	6	=	6
2	x	6	=	12
3	x	6	=	18
4	x	6	=	24
5	x	6	=	30
6	x	6	=	36
7	x	6	=	42
8	x	6	=	48
9	x	6	=	54
10	x	6	=	60
11	x	6	=	66
12	x	6	=	72

7 x table

1	x	7	=	7
2	x	7	=	14
3	x	7	=	21
4	x	7	=	28
5	x	7	=	35
6	x	7	=	42
7	x	7	=	49
8	x	7	=	56
9	x	7	=	63
10	x	7	=	70
11	x	7	=	77
12	x	7	=	84

8 x table

1	x	8	=	8
2	x	8	=	16
3	x	8	=	24
4	x	8	=	32
5	x	8	=	40
6	x	8	=	48
7	x	8	=	56
8	x	8	=	64
9	x	8	=	72
10	x	8	=	80
11	x	8	=	88
12	x	8	=	96

9 x table

1	x	9	=	9
2	x	9	=	18
3	x	9	=	27
4	x	9	=	36
5	x	9	=	45
6	x	9	=	54
7	x	9	=	63
8	x	9	=	72
9	x	9	=	81
10	x	9	=	90
11	x	9	=	99
12	x	9	=	108

10 x table

1	x	10	=	10
2	x	10	=	20
3	x	10	=	30
4	x	10	=	40
5	x	10	=	50
6	x	10	=	60
7	x	10	=	70
8	x	10	=	80
9	x	10	=	90
10	x	10	=	100
11	x	10	=	110
12	x	10	=	120

11 x table

1	x	11	=	11
2	x	11	=	22
3	x	11	=	33
4	x	11	=	44
5	x	11	=	55
6	x	11	=	66
7	x	11	=	77
8	x	11	=	88
9	x	11	=	99
10	x	11	=	110
11	x	11	=	121
12	x	11	=	132

12 x table

1	x	12	=	12
2	x	12	=	24
3	x	12	=	36
4	x	12	=	48
5	x	12	=	60
6	x	12	=	72
7	x	12	=	84
8	x	12	=	96
9	x	12	=	108
10	x	12	=	120
11	x	12	=	132
12	x	12	=	144

9 Use fractions, decimals and percentages to describe situations

Calculate with fractions, decimals, percentages and ratios

Before you can use fractions, decimals and percentages to describe situations, you need to have a thorough understanding of all three and the relationship between them.

Fractions

A fraction is a small piece or any part of a whole. It has two main parts: a numerator and a denominator. You can remember it by the acronym NOD.

Numerator
Over
Denominator

The *denominator* tells you how many parts the whole has been divided up into and the *numerator* tells you how many parts of the whole are actually used. For example, a numerator 3 and a denominator 8 will give us $\frac{3}{8}$, as represented in Fig 9.1.

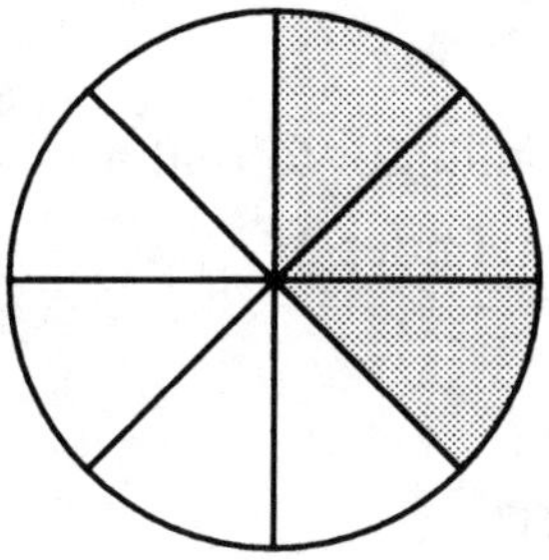

Fig 9.1 Representation of $\frac{3}{8}$

■ Types of fractions

There are three types of fractions: a proper fraction, an improper fraction and a mixed number.

- A *proper fraction* has the numerator smaller than the denominator and therefore is always less than 1: e.g. $\frac{3}{8}$.
- An *improper fraction* has the numerator larger than the denominator and is therefore always greater than 1: e.g. $\frac{12}{5}$.
- A *mixed number* is a mixture of whole numbers and a proper fraction and is therefore always greater than 1: e.g. $2\frac{2}{5}$.

An improper fraction can be changed to a mixed number and a mixed number can be changed to an improper fraction: e.g. $\frac{12}{5}$ is the same as $2\frac{2}{5}$.

EXAMPLE 1

(a) Convert $2\frac{3}{4}$ into an improper fraction.

The denominator tells you a whole number is to be divided into 4 parts.
Two whole numbers with 4 parts in each equals 8 parts ($2 \times 4 = 8$).
Add the 3 parts from the fraction to the 8 parts from the whole ($3 + 8 = 11$).

Answer = $\frac{11}{4}$

(b) Convert $\frac{14}{3}$ into a mixed number

Divide the numerator by the denominator to find how many whole numbers there are ($14 \div 3 = 4$, remainder 2)

Answer = $4\frac{2}{3}$.

Fractions in their lowest form

It is always best to display fractions in their lowest form: e.g. $\frac{4}{8}$ becomes $\frac{1}{2}$ in its lowest form. You must divide both numerator and denominator by the same number. The most common numbers to look for are multiples of 2, 3, 5 or 10.

EXAMPLE 2

Cancel $\frac{16}{48}$ down into its lowest form.

Divide both the numerator and the denominator by 16 ($16 \div 16 = 1$, $48 \div 16 = 3$).

Answer = $\frac{1}{3}$.

Addition and subtraction of fractions

You need to be able to add and subtract fractions; this should preferably be without the use of a fraction button on a calculator.

- When adding or subtracting fractions that have the same denominator you simply add or subtract the numerators.
- When adding or subtracting fractions with different denominators you have to find a lowest common denominator, LCD. This is the lowest number into which both denominators will divide. Whatever number you multiply the denominator by to get the LCD, you must multiply the numerator by the same number.

EXAMPLE 3

(a) $\frac{3}{8}+\frac{2}{8}=\frac{5}{8}$

(b) $\frac{2}{3}-\frac{1}{6}=\frac{4}{6}-\frac{1}{6}=\frac{3}{6}=\frac{1}{2}$

6 is the LCD; it can be divided by itself and also by 3.

(c) $\frac{1}{3}+\frac{2}{5}$

15 is the lowest number that both 3 and 5 will divide into.

$$\frac{1}{3}+\frac{2}{5} = \frac{(1\times 5)}{(3\times 5)}+\frac{(2\times 3)}{(5\times 3)} = \frac{5}{15}+\frac{6}{15} = \frac{11}{15}$$

Multiplication and division of fractions

You need to be able to multiply and divide fractions.

- To multiply fractions, you multiply the numerators together and multiply the denominators together.
- To divide fractions, you have to invert (turn upside down) the right-hand fraction and then multiply the fractions together.

EXAMPLE 4

(a) $\frac{2}{3}\times\frac{4}{5}=\frac{8}{15}$

(b) $\frac{2}{7}\div\frac{1}{3} = \frac{2}{7}\times\frac{3}{1} = \frac{6}{7}$

When multiplying or dividing mixed fractions, it is always advisable to convert the fractions to improper fractions. This saves making unnecessary mistakes.

Exercise 9.1

1 Convert the following mixed numbers into improper fractions:

(a) $3\frac{1}{2}$ (b) $4\frac{3}{5}$ (c) $5\frac{1}{4}$ (d) $2\frac{8}{11}$ (e) $3\frac{4}{7}$

2 Convert the following improper fractions into mixed numbers:

(a) $\frac{11}{4}$ (b) $\frac{22}{3}$ (c) $\frac{19}{8}$ (d) $\frac{17}{5}$ (e) $\frac{23}{6}$

3 Cancel down the following into their lowest form:

(a) $\frac{8}{32}$ (b) $\frac{24}{48}$ (c) $\frac{28}{36}$ (d) $\frac{77}{112}$ (e) $\frac{102}{132}$

4 Add the following fractions:

(a) $\frac{2}{3} + \frac{1}{4}$ (b) $\frac{1}{2} + \frac{5}{6}$ (c) $\frac{3}{4} + \frac{4}{5}$ (d) $\frac{1}{3} + \frac{2}{5} + \frac{1}{2}$ (e) $\frac{1}{4} + \frac{1}{6} + \frac{2}{3}$

5 Subtract the following fractions:

(a) $\frac{3}{8} - \frac{1}{4}$ (b) $\frac{11}{16} - \frac{5}{8}$ (c) $\frac{3}{5} - \frac{1}{7}$ (d) $\frac{5}{6} - \frac{4}{9}$ (e) $\frac{5}{8} - \frac{7}{16}$

6 Multiply the following fractions:

(a) $\frac{1}{2} \times \frac{4}{13}$ (b) $\frac{4}{7} \times \frac{4}{9}$ (c) $\frac{12}{17} \times \frac{2}{3}$ (d) $\frac{9}{20} \times \frac{25}{27}$ (e) $\frac{17}{22} \times \frac{11}{34}$

7 Divide the following fractions:

(a) $\frac{3}{8} \div \frac{5}{8}$ (b) $\frac{7}{16} \div \frac{1}{4}$ (c) $\frac{13}{21} \div \frac{3}{7}$ (d) $\frac{15}{18} \div \frac{13}{12}$ (e) $\frac{4}{9} \div \frac{24}{27}$

Decimals

Whole numbers and decimals are constructed as shown in Fig 9.2.

Decimal point ↓

1,000's	100's	10's	Units	•	$\frac{1}{10}$	$\frac{1}{100}$	$\frac{1}{1000}$

Fig 9.2

If we put 0.235 into the table, we obtain the result shown in Fig 9.3.

1,000's	100's	10's	Units	•	$\frac{1}{10}$	$\frac{1}{100}$	$\frac{1}{1000}$
			0	•	**2**	**3**	**5**

Fig 9.3

$$0 \text{ units plus } \frac{2}{10} \text{ plus } \frac{3}{100} \text{ plus } \frac{5}{1000}$$

or

$$\frac{235}{1000}$$

As you can see, a decimal is a fraction, but it is written without a denominator.

■ Adding and subtracting decimals

The golden rule to remember when adding and subtracting decimals on paper is to *always* line up the decimal points and numbers in columns.

EXAMPLE 5

(a) 0.25 + 2.375 + 73.425 + 137.5

```
   0.25
   2.375
  73.425
 137.5
 _______
 213.550
 _______
```

(b) 967.825 – 63.25

```
 967.825
  63.25
 _______
 904.575
 _______
```

If you use graph paper when adding and subtracting decimals, the squares will help you get all the numbers in the correct columns.

■ Multiplying decimals

When multiplying two decimal numbers:

1 Count the total number of decimal places in the numbers being multiplied.
2 Disregard the decimal points when multiplying the two numbers.
3 Replace the decimal point in the product by counting the places from right to left.

EXAMPLE 6

Multiply 3.25 by 5.2.

Total number of decimal places is 3. Ignore the decimal points.

$$\begin{array}{r} 325 \\ 52 \\ \hline 16900 \\ \hline \end{array}$$

Replace the decimal point (3 places from right).

Answer = 16.900

■ Dividing decimals

When dividing one decimal number by another:

1 Make the divisor into a whole number. Do this by moving the decimal point from left to right.
2 What you do to one number, you must do to the other. Move the decimal point in the number being divided the same number of places from left to right.
3 Divide in the usual way.

EXAMPLE 7

Divide 4.505 by 0.85.

0.85 becomes 85	(decimal point moves two places to right)
4.505 becomes 450.5	(this decimal point must also move two places to the right)

$$85 \overline{)\ 450.5}\quad = 5.3$$

Answer = 5.3

Exercise 9.2

Do not use a calculator.

1 Add the following decimals:

(a) 0.969 + 0.661
(b) 208.16 + 23.719
(c) 5.74 + 39.3 + 2.63
(d) 4374.46 + 25.2 + 675.06
(e) 0.434 + 64.8 + 744.2 + 10.3

2 Subtract the following decimals:

(a) 457.87 – 80.07
(b) 27.4 – 6.3
(c) 731.649 – 210.23
(d) 50889.51 – 8424.11
(e) 78189.519 – 38836.017

3 Multiply the following decimals:

(a) 47.6×0.3
(b) 361.3×7.2
(c) 0.518×34.6
(d) 0.048×0.178
(e) 2919.96×8.35

4 Divide the following decimals:

(a) $0.924 \div 0.7$
(b) $17.6 \div 0.02$
(c) $696.6 \div 12.9$
(d) $142.272 \div 31.2$
(e) $13.9854 \div 0.22$

Percentages

A percentage is a fraction with a denominator of 100: e.g. 33% is $\frac{33}{100}$. It is a convention to use the % sign instead of the denominator.

- To change a percentage into a fraction or decimal, divide it by 100.
- To change a fraction or decimal into a percentage, multiply it by 100.

EXAMPLE 8

(a) Express 75% as a fraction.
Answer = $\frac{75}{100}$
This will cancel down to $\frac{3}{4}$.

(b) Express 6% as a decimal.
$6 \div 100$
Answer = 0.06

(c) Express $\frac{8}{25}$ as a percentage.
$\frac{8}{25} \times 100$
Answer = 32%

(d) Express 0.075 as a percentage
0.075×100
Answer = $7\frac{1}{2}$%

To find the percentage of a quantity, you must first convert the percentage into a fraction or a decimal number. When you have done that you multiply by the quantity.

EXAMPLE 9

(a) Calculate 30% of 80 grams.
Convert to a decimal number.

$30\% = 30 \div 100 = 0.3$
$0.3 \times 80 = 24$
Answer = 24 grams

(b) Calculate 15% of £220.
Convert to a fraction.

$$15\% = \frac{15}{100} = \frac{3}{20}$$

$$\frac{3}{20} \times 220 = \frac{660}{20} = 33$$

Answer = £33

(c) A football team won 23 games out of 40. What percentage is this?
23 out of 40 is a fraction.

$$\frac{23}{40} \times 100 = 57\frac{1}{2}\%$$

(d) 20% of a length of rope measures 600 cm. What is the complete length?

20% as a decimal = 0.2
Let the complete length be x. Therefore 20% of x = 600 cm.

$$0.2 \times x = 600$$
$$x = 600 \div 0.2 = 3000$$

Answer = 3000 cm

The performance criteria analysis is shown in Fig 9.4 (on page 64).

Exercise 9.3

1 Calculate:

(a) 5% of 5420
(b) 20% of 380
(c) 80% of 400 g
(d) 67% of 13 500 m
(e) 42% of £415

2 Calculate:

(a) 105% of 520
(b) 220% of 30
(c) 180% of 400 kg
(d) 6% of 1500 km
(e) $\frac{1}{2}$% of £120

3 Calculate the percentage for:

(a) 25 out of 125
(b) 5 out of 20
(c) 45 out of 375
(d) 568 g out of 1704 g
(e) 356 people out of 800 people

4 Calculate the value for x if:

(a) 25% of x is 250
(b) 55% of x is 4125
(c) 47% of £x is £40.42
(d) $12\frac{1}{2}$% of x km is 5948 km
(e) 66.7% of x ml is 6003 ml

	Performance Criteria	Pass or NYA
1	techniques appropriate to the task are selected and used by the individual	Pass Calculated Percentage
2	activities required by the techniques are performed to appropriate levels of precision and in the correct sequence	Pass. converted to a decimal or fraction first
3	mathematical terms and mathematical aspects of everyday language are used and interpreted with precision	Pass interpreted the verb "to Calculate"
4	use estimation to check calculations	NYA Did not estimate
5	calculations and results obtained are correct	Pass Correct answer
6	aids are used correctly	Pass used Calculator correctly
7	conclusions / generalisations / predictions drawn from results are valid	NYA none drawn
8	clear justification is given for the interpretation of results	NYA no interpretation

Fig 9.4

Ratios

A ratio is a proportion. Cement mortar is mixed in a proportion of 1 part cement to 3 parts sand and is written as a ratio 1:3. It does not matter what is used to measure the cement and sand as long as they both are measured by the same item, e.g. 1 wheelbarrow of cement and 3 wheelbarrows of sand, or 1 bag of cement and 3 bags of sand. Concrete for paths needs a proportion of 1 part cement, 2 parts sand and 3 parts coarse aggregate: a ratio of 1:2:3 (1 wheelbarrow of cement, 2 wheelbarrows of sand and 3 wheelbarrows of coarse aggregate).

Ratios should be treated the same way as fractions. What you do to one side, you must do to the other. A large ratio can be cancelled down into its simplest term: e.g. 8:4 is the same as 2:1 because both sides are divisible by 4.

■ Calculating a ratio

To calculate a ratio, you must:

1. Calculate the total proportional parts, e.g. 2:3 has 5 total parts (2 + 3 = 5).
2. Divide the quantity by the total proportional parts, e.g. 10 g ÷ 5 = 2 g per part.
3. Multiply the quantity per part by each of the ratios, e.g. 2:3 is 2 × 2 g : 3 × 2 g = 4 g : 6 g.

EXAMPLE 10

(a) Divide 500 g of sugar in the ratio 7:3.

Total parts	7 + 3 = 10
Quantity per part	500 g ÷ 10 = 50 g
Multiply out	7 × 50 g : 3 × 50 g = 350 g : 150 g

Answer = 350 g:150 g

(b) Divide £275 in the ratio 10:8:7.

Total parts	$10 + 8 + 7 = 25$
Quantity per part	$£275 \div 25 = £11$
Multiply out	$10 \times £11 : 8 \times £11 : 7 \times £11 = £110 : £88 : £77$

Answer = £110 : £88 : £77

You can double check your answer by adding the £'s together.

£110 + £88 + £77 = £275

The performance criteria analysis is shown in Fig 9.5.

	Performance Criteria	Pass or NYA
1	techniques appropriate to the task are selected and used by the individual	Pass. used ratio technique
2	activities required by the techniques are performed to appropriate levels of precision and in the correct sequence	Pass. found part then calculated
3	mathematical terms and mathematical aspects of everyday language are used and interpreted with precision	Pass. Used divide correctly
4	use estimation to check calculations	NYA no estimate
5	calculations and results obtained are correct	Pass correct answer
6	aids are used correctly	Pass used calculator correctly
7	conclusions / generalisations / predictions drawn from results are valid	NYA Did not do
8	clear justification is given for the interpretation of results	Pass. Double checked answer

Fig 9.5

Exercise 9.4

1 Divide the following into the ratios shown:

(a) £350 in the ratio 2:3
(b) 1000 km in the ratio 13:12
(c) 155 g in the ratio 3:2
(d) 660 cm in the ratio 19:11
(e) 282 ml in the ratio 27:20

2 Divide the following into the ratios shown:

(a) 700 g in the ratio 2:3:2
(b) 243 km in the ratio 11:9:7
(c) 128 litres in the ratio 8:5:3
(d) £5946 in the ratio 2:1:3
(e) 1.5 m in the ratio 17:9:4

Use fractions, decimals and percentages to describe situations

Any of the three can be used when describing the results of a survey or questionnaire.

EXAMPLE 11

Forty people were interviewed about a certain washing powder. Twenty-four people use it all the time, 6 people use it occasionally and 10 people never use it.

(a) Express 24 out of 40 as a fraction

$$\frac{24}{40} = \frac{3}{5}$$

(b) Express 6 out of 40 as a decimal

$$6 \div 40 = 0.15$$

(c) Express 10 out of 40 as a percentage.

$$\frac{10}{40} \times 100 = 25\%$$

Answers = (a) $\frac{3}{5}$ people interviewed use the washing powder all the time.

(b) 0.15 people interviewed use the washing powder occasionally.

(c) 25% people interviewed never use the washing powder.

Figure 9.6 shows the performance criteria analysis for this example.

	Performance Criteria	Pass or NYA
1	techniques appropriate to the task are selected and used by the individual	Pass. Calculated percentage, fraction and decimal
2	activities required by the techniques are performed to appropriate levels of precision and in the correct sequence	Pass. Calculated correctly
3	mathematical terms and mathematical aspects of everyday language are used and interpreted with precision	Pass. understood "interviewed"
4	use estimation to check calculations	NYA did not estimate
5	calculations and results obtained are correct	Pass all answers correct
6	aids are used correctly	Pass used calculator correctly
7	conclusions / generalisations / predictions drawn from results are valid	Pass. Results are valid, based upon interviews
8	clear justification is given for the interpretation of results	Pass. shown working out and handed in interview sheets

Fig 9.6

Exercise 9.5

1 The price of a train ticket goes up from £28 to £35. What is the percentage increase?

2 A theatre sold 375 tickets for an evening performance. If the theatre seats 600, what fraction is occupied?

3 A company has a budget for redecoration of £10 000, but spends £10 700. By what percentage did it overspend?

4 500 employees were surveyed about the standard of hygiene in the staff canteen. 375 employees thought the standard to be above average. Express this as a decimal.

5 12 employees out of 76 were consistently late for work. 54 employees were never late for work and the remainder were only occasionally late. What percentage of employees are (a) consistently late; (b) never late?

Calculate with fractions, decimals, percentages and ratios

The exercises in the last section have given you the opportunity to practise calculation with numbers. Ideally you should calculate with fractions, decimals, percentages and ratios to solve problems.

Exercise 9.6

1 The rent of a flat costs £35 per week. If the cost is increased by 9%, what is the new rental charge?

2 An employee receives a Christmas bonus of £90. It represents $\frac{6}{835}$ of his annual salary. Calculate his annual salary.

3 A map has a scale of 2.5 cm:5 km. The measurement, on the map, between two towns is 9 cm. What is the distance in km?

4 A store offers a cash discount of 12%. If £365 worth of goods are bought with cash, how much will be saved?

5 Three employees are to share a bonus of £1000. Their annual salaries are in a ratio of 13:11:8 and the bonus is to be paid in the same ratio. How much will each receive?

6 A company employs 500 people. $\frac{7}{10}$ of the workforce are female. How many female employees does the company have?

7 A building firm has a contract to increase the size of the local library by 3.5 times. If the present buildings are 37 500 m^3, what will be the size of the extension?

8 A store sells two brands of peas. Brand A weighs 230 g and costs 35p and Brand B weighs 250 g and costs 38p. Which brand is better value for money?

9 A children's home spends $\frac{1}{2}$ of its income on food, $\frac{1}{3}$ on clothes and $\frac{1}{6}$ on electricity. If the income for the year was £24 120, how much was spent on each of the three items?

10 A greengrocer received a delivery of fruit, of which 8% was rotten. If he was able to sell 552 kg, what was the original weight of fruit?

10 Find perimeters, areas and volumes

Find areas of plane shapes and volumes of simple solids

Perimeter

Perimeter is the distance around a shape. You need to know how to calculate the perimeter of a square, a rectangle and a triangle.

EXAMPLE 1

Calculate the perimeter of the badminton court shown in Fig 10.1 using imperial units.

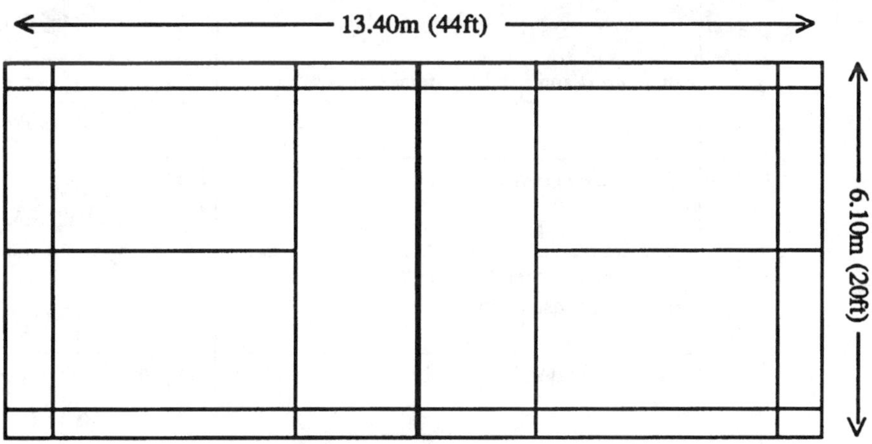

Fig 10.1 A badminton court

The perimeter of a rectangle is:

$$(2 \times \text{length}) + (2 \times \text{width})$$

or

$$\text{length} + \text{width} + \text{length} + \text{width}$$

Therefore the perimeter of the badminton court in Fig 10.1 is:

$$(2 \times 44 \text{ ft}) + (2 \times 20 \text{ ft}) = 128 \text{ ft}$$

or

$$44 \text{ ft} + 20 \text{ ft} + 44 \text{ ft} + 20 \text{ ft} = 128 \text{ ft}$$

The performance criteria analysis for this example is shown in Fig 10.2.

	Performance Criteria	Pass or NYA
1	techniques appropriate to the task are selected and used by the individual	Pass 2L+2w
2	activities required by the techniques are performed to appropriate levels of precision and in the correct sequence	Pass. Added lengths of sides
3	mathematical terms and mathematical aspects of everyday language are used and interpreted with precision	Pass. Perimeter formula used
4	use estimation to check calculations	NYA no estimate
5	calculations and results obtained are correct	Pass correct answer
6	aids are used correctly	Pass Calculator
7	conclusions / generalisations / predictions drawn from results are valid	NYA maybe not applicable
8	clear justification is given for the interpretation of results	NYA. maybe not applicable

Fig 10.2

Exercise 10.1

1 Find the perimeter of each shape in Fig 10.3.

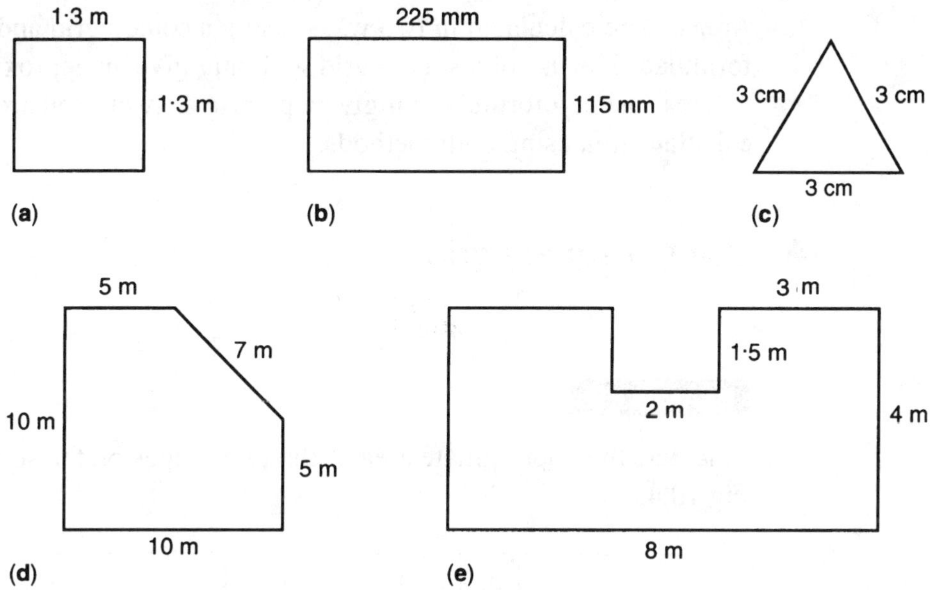

Fig 10.3

2 A lounge is 5 m × 3 m with a chimney breast of 0.5 m deep by 1.5 m wide. The owner has contracted you to put up a 15 cm wide wall-paper border around the room. What length of border will you require?

3 A farmer wishes to build a wire fence around a field 2206 m × 3673 m. How much fencing will the farmer require?

4 A long distance runner wishes to use the local park for week-day training. If the runner wishes to run 6000 m per day and the park is 940 m × 560 m, how many circuits of the park must be completed to reach this target?

5 A triangular bandage needs a border to stop the bandage from fraying. How much border is needed if the bandage measures 1200 mm × 900 mm × 900 mm?

Area of plane shapes

The area of a plane shape is measured by the number of square units it contains.

Area can be calculated in two ways; using a square grid and using formulae. The use of a square grid will only give an approximate answer whereas using a formula will give a precise answer. You are required to calculate area using both methods.

■ Using a square grid

EXAMPLE 2

Calculate the approximate area of the two shapes on the square grid in Fig 10.4.

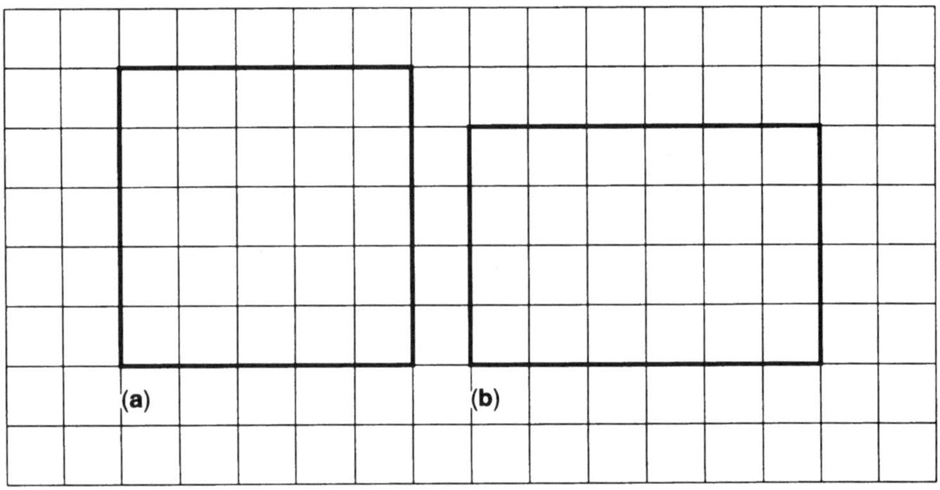

Fig 10.4

To calculate the area, count the number of squares within the shape. For the shapes in Fig 10.4:

The area of **a** = 25 squares
The area of **b** = 24 squares.

Exercise 10.2

Calculate an approximate area for the shapes in Fig 10.5.

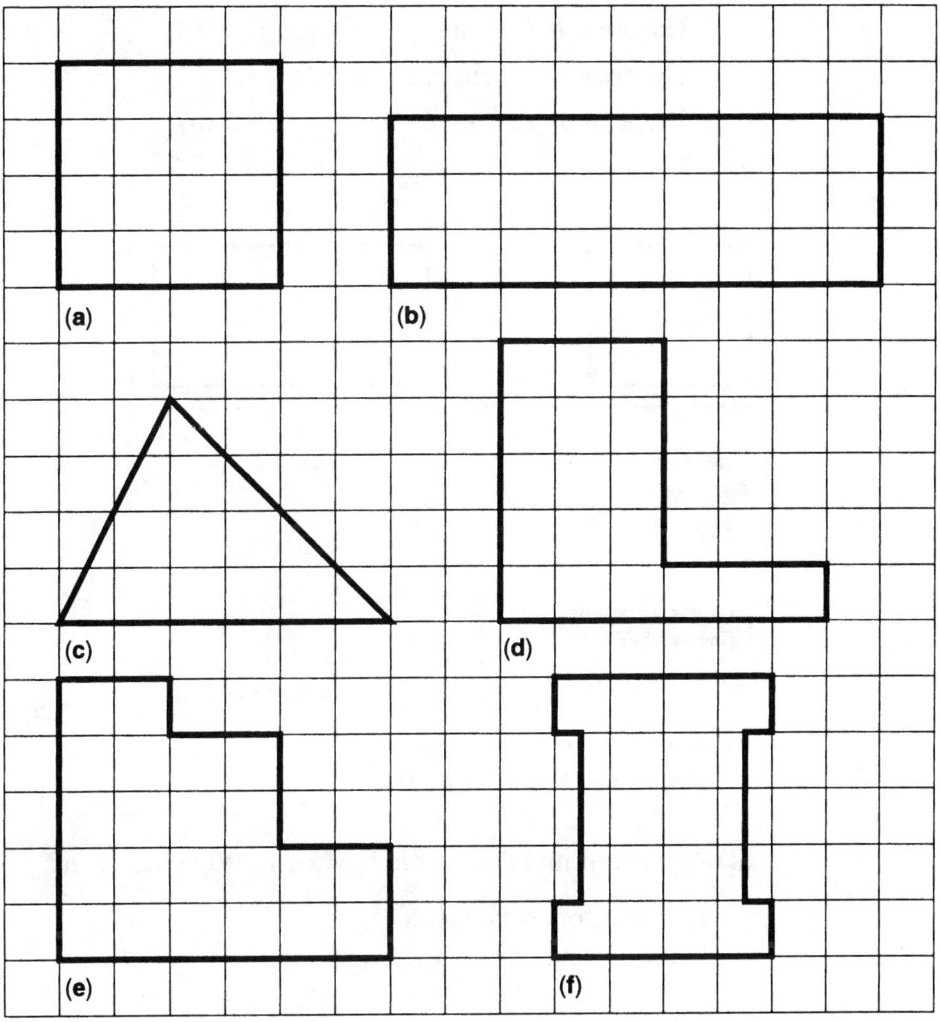

Fig 10.5

(Scale: 1 square = 1 cm^2)

Using formulae

Look at the shapes in Fig 10.6 and learn the following formulae for calculating their areas:

The area of a square	=	$l \times b$
The area of a rectangle	=	$l \times b$
The area of a triangle	=	$\frac{1}{2} b \times h$

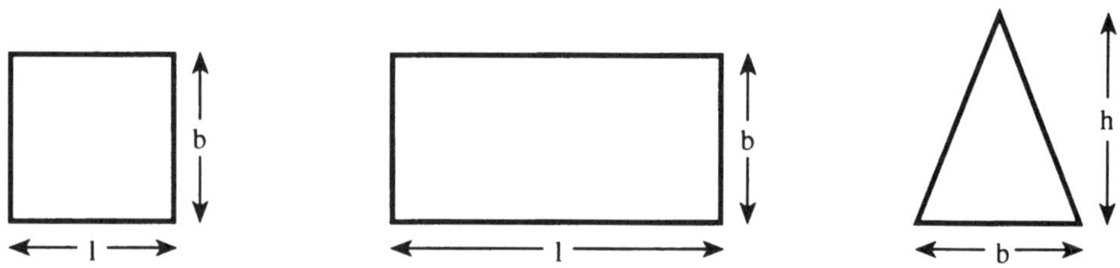

(a) $l \times b$ **(b)** $l \times b$ **(c)** $\frac{1}{2} b \times h$

Fig 10.6

EXAMPLE 3

(a) A carpet is 3 m square. What area is it?

Formula for square is $l \times b$.
Answer = 3 m × 3 m = 9 m^2.

(b) A sheet of metal is 20 cm by 15 cm. What area is it?

Formula for rectangle is $l \times b$.
Answer = 20 cm × 15 cm = 300 cm^2.

(c) A triangular sign has a base of 60 cm and a height of 70 cm. What area is it?

Formula for a triangle is $\frac{1}{2} b \times h$.

Answer = $\frac{1}{2}$ × 60 cm × 70 cm = 2100 cm^2.

The performance criteria analysis for this example appears in Fig 10.7.

	Performance Criteria	Pass or NYA
1	techniques appropriate to the task are selected and used by the individual	Pass. used formulae for square, rectangle and triangle
2	activities required by the techniques are performed to appropriate levels of precision and in the correct sequence	Pass. multiplied numbers out
3	mathematical terms and mathematical aspects of everyday language are used and interpreted with precision	Pass. calculated area
4	use estimation to check calculations	NYA
5	calculations and results obtained are correct	Pass ✓
6	aids are used correctly	Pass calculator
7	conclusions / generalisations / predictions drawn from results are valid	NYA Maybe not applicable
8	clear justification is given for the interpretation of results	NYA Maybe not applicable

Fig 10.7

Exercise 10.3

1 A garden is 5 m × 7 m. What area is it?

2 What is the area of a sheet of metal measuring 58 cm × 32 cm?

3 Find the area of a field measuring 507 m × 210 m.

4 Find the area of a triangular bandage that has a base of 1200 mm and a perpendicular height of 600 mm.

5 A triangular-shaped clock has a base of 60 cm and a perpendicular height of 70 cm. How much area will it take up on an office wall?

6 The stage of a local theatre measures 45 m by 30 m. A travelling theatre group needs 1200 m^2 for its performance. Is the stage large enough?

7 The blood bank wishes to use a community centre for a blood donor session. Trellis Hall is 5 m by 9 m and Cumberland Hall is 7 m by 8 m. The blood bank has five beds, each bed requiring 6 m^2 of floor area. Which hall would be most suitable?

8 The hold of a cross channel ferry measures 25.2 m by 153 m and cars are parked end to end. If the average car takes up an area of 16.20 m^2, how many cars will the ferry carry?

9 A confectionery company is selling its chocolates in a triangular based box. The base measurements are given as 7.5 cm and 6.5 cm. If a shop allocated 195 cm^2 of shelf space, how many boxes can be displayed at a time?

10 A company makes *Give Way* signs. Each sign has a base of 0.6 m and a height of 0.6 m. How many signs can be made out of 2 m^2 of metal?

Volumes of simple solids

The volume of a simple solid is measured by the number of cubic units it contains. You are required to know the formulae for a cube (or cuboid) and a prism. (See Fig 10.8.)

The volume of a cube $= l \times b \times h$

The volume of a cuboid $= l \times b \times h$

The volume of a prism $= \frac{1}{2} b \times h \times l$ (or area of cross-section $\times\ l$)

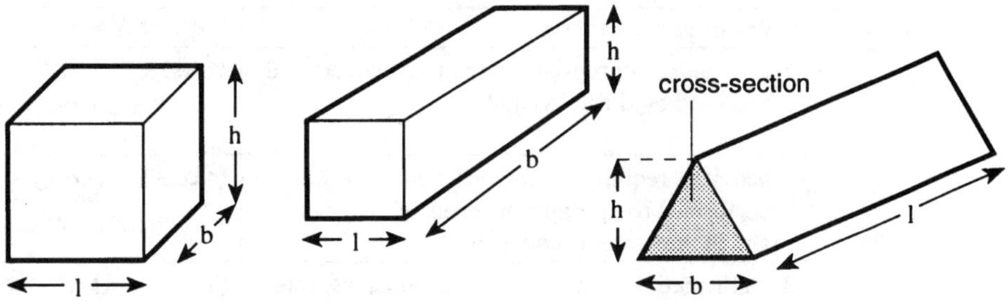

(a) $l \times b \times h$ **(b)** $l \times b \times h$ **(c)** $\frac{1}{2} b \times h \times l$

Fig 10.8 ***(or area of cross-section × l)***

EXAMPLE 4

(a) A box is a 3 m cube. What volume is it?

The length, width and height of a cube are the same measurement. The formula for the volume of a cube is:

$l \times b \times h$

Answer = 3 m × 3 m × 3 m = 27 m^3.

(b) A metal container is 3 m by 6 m by 11 m. What volume is it? The length, width and height of a cuboid are different measurements. (Two measurements may be the same.) The formula for the volume of a cuboid is:

$l \times b \times h$

Answer = 3 m × 6 m × 11 m = 198 m^3.

(c) A prism has a base of 4 cm, a height of 4 cm and a length of 15 cm. What volume is it?
The formula for the volume of a prism is:

half the base × the perpendicular height × length

or

the area of cross-section × length

Answer = 2 cm × 4 cm × 15 cm = 120 cm^3

The performance criteria analysis for this example appears in Fig 10.9.

	Performance Criteria	Pass or NYA
1	techniques appropriate to the task are selected and used by the individual	Pass volume formula
2	activities required by the techniques are performed to appropriate levels of precision and in the correct sequence	Pass multiplied out numbers
3	mathematical terms and mathematical aspects of everyday language are used and interpreted with precision	Pass volume formulae used precisely
4	use estimation to check calculations	NYA no estimate
5	calculations and results obtained are correct	Pass ✓
6	aids are used correctly	N/A no aid used.
7	conclusions / generalisations / predictions drawn from results are valid	N/A
8	clear justification is given for the interpretation of results	N/A

Fig 10.9

Exercise 10.4

1 A dice is 20 mm × 20 mm × 20 mm. What is its volume?

2 What is the volume of a container lorry that measures 3.5 m × 4.3 m × 11.2 m?

3 Find the volume of a filing cabinet that measures 47 cm × 62 cm × 132 cm.

4 Find the volume of a prism shaped piece of metal. The length measures 3000 mm and the area of the cross-section measures 2550 mm^2.

5 Find the volume of a triangular shaped box of chocolates that has a base of 30 cm, a height of 17 cm and a length of 5 cm.

6 An office replaces five filing cabinets measuring 50 cm × 60 cm × 140 cm with three roll front cabinets measuring 110 cm × 50 cm × 200 cm. How much more space will the new cabinets require?

7 A storage company own 320 containers and needs to buy a new warehouse. Each storage container measures 4.6 m × 3.2 m × 8.7 m. What is the minimum-size warehouse the company should buy? Give your answer in m^3.

8 A jeweller has 2430 mm^3 of gold that he wishes to make into charms for a bracelet. If each charm has a length of 4 mm and a cross-sectional area of 4.5 mm^2, how many charms will he make?

9 A box 30 cm × 32 cm × 57 cm contains 14 books. What is the volume of each book?

10 From a cuboid piece of stone 60 cm × 60 cm × 90 cm a mason carves a prism-shaped table support. The support has a base of 60 cm, a height of 60 cm and a length of 90 cm. What volume of stone has been wasted?

Calculating area and volume with mixed units

The rule to remember when calculating with mixed units is always to convert to the units required in the answer before calculating area and volume.

EXAMPLE 5

(a) A carpet measures 95 cm × 3 m. What is its area in m^2?

95 cm = 0.95 m

Answer = 0.95 m × 3 m = 2.85 m^2.

A common mistake is to convert to cm (Answer = 28 500 cm^2) and then divide by 100. There are 100 cm in a metre, but 10 000 cm^2 in 1 m^2.

$$1 \text{ m} \times 1 \text{ m} = 1 \text{ m}^2$$

This is the same as

$$100 \text{ cm} \times 100 \text{ cm} = 10\,000 \text{ cm}^2$$

Therefore if 1 m^2 = 10 000 cm^2, 28 500 cm^2 = 2.85 m^2

(b) A wooden box measures 50 cm × 70 cm × 2 m. What is the volume in cubic metres?

50 cm = 0.5 m
70 cm = 0.7 m

Answer = 0.5 m × 0.7 m × 2 m = 0.7 m^3

The mistake mentioned for area is also made with volume. Many people divide a cm^3 answer by 100, instead of 1 000 000.

$$50 \text{ cm} \times 70 \text{ cm} \times 200 \text{ cm} = 700\,000 \text{ cm}^3$$

To express answer in m^3, remember:

$$1 \text{ m} \times 1 \text{ m} \times 1 \text{ m} = 1 \text{ m}^3$$

This is the same as

$$100 \text{ cm} \times 100 \text{ cm} \times 100 \text{ cm} = 1\,000\,000 \text{ cm}^3$$

Therefore if 1 m^3 = 1 000 000 cm^3, 700 000 cm^3 = 0.7 m^3.

11 Know and use the formulae for finding the area and circumference of circles

It is not sufficient just to know the formulae when calculating the area and circumference of circles – you must understand and know what each part of the formula means.

The two formulae are:

Circumference	Pi times the diameter	πd
Area	Pi times the square radius	πr^2

The *circumference* is the distance around the edge of a circle. *Pi* is the ratio of the circumference of a circle to its diameter and equals 3.142 to three decimal points (3 dp) or $\frac{22}{7}$. This means that the circumference is just over three times longer than the diameter. The *diameter* is the distance across the centre of a circle from one edge to the other. The *radius* is the distance from the centre of a circle to the outer edge. Therefore, the radius is half the length of the diameter (see Fig 11.1).

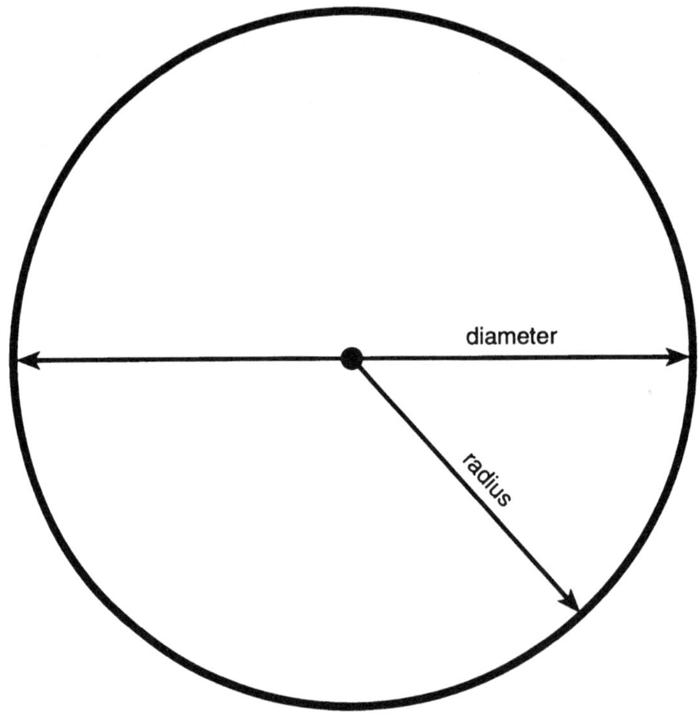

Fig 11.1 The radius and diameter of a circle

Circumference of a circle

EXAMPLE 1

(a) A circle has a diameter of 6 cm. What is the circumference?

Formula is πd.

Answer = 3.142×6 cm = 18.852 cm.

If you use the pi (π) button on a scientific calculator, the answer will be 18.849556 cm or 18.85 cm (2 dp).

(b) What is the circumference of a circle that has a radius of 2.5 cm?

Formula is πd. Diameter is twice the length of the radius.

Answer = $3.142 \times 2 \times 2.5$ cm = 15.71 cm.

The pi button on a calculator gives an answer of 15.707963 cm or 15.71 cm (2 dp).

(c) A circle has a circumference of 22 m. What is the diameter?

$$3.142 \times d = 22 \text{ m}$$

$$d = \frac{22}{3.142} \text{ m}$$

$$d = 7 \text{ m}$$

It is not sufficient to think you can rely on a calculator to remember pi for you. You should make an effort to remember that pi = 3.142.

Area of a circle

EXAMPLE 2

(a) A circle has a radius of 3 m. What is its area?

Formula is πr^2.

$3.142 \times 3^2 = 28.26$

or

$3.142 \times 3 \times 3 = 28.26$

Area = 28.26 m^2.

(b) What is the area of a circle that has a diameter of 8 cm?

Formula is πr^2. Diameter divided by 2 = radius

$8 \div 2 = 4$

The radius is therefore 4 cm.

$3.142 \times 4^2 = 50.272$

Area = 50.272 cm^2.

(c) What is the radius of a circle if the area is 78.55 cm^2.

$$3.142 \times r^2 = 78.55$$
$$r^2 = \frac{78.55}{3.142}$$
$$r^2 = 25$$
$$r = \sqrt{25}$$
$$r = 5$$

Radius = 5 cm.

The performance criteria analysis for this example appears in Fig 11.2.

	Performance Criteria	Pass or NYA
1	techniques appropriate to the task are selected and used by the individual	Pass used πd and πr^2
2	activities required by the techniques are performed to appropriate levels of precision and in the correct sequence	Pass multiplied through
3	mathematical terms and mathematical aspects of everyday language are used and interpreted with precision	Pass π interpreted as 3.142
4	use estimation to check calculations	NYA
5	calculations and results obtained are correct	Pass Correct answer
6	aids are used correctly	Pass Calculator
7	conclusions / generalisations / predictions drawn from results are valid	N/A
8	clear justification is given for the interpretation of results	N/A

Fig 11.2

Exercise 11.1

1 If the diameter of a circle is 30 mm, what is its circumference?

2 What is the area of a circle that has a radius of 5 cm?

3 Find the circumference and area of a circle that has a radius of 65 mm.

4 The area of a circle is 50.272 m^2. What is the radius?

5 Find the diameter of a circle if the circumference is 9.426 cm.

6 The diameter of a car wheel is 57 cm. What is the area?

7 A young mother wishes to erect a fence around a garden pond to protect her children. If the pond is 2.5 m in diameter, how much fence will she need?

8 A Town Hall clock has a long hand that measures 75 cm. What are the circumference and area of the clock?

9 A builder knows the circumference of a piece of pipe to be 47 130 mm. What is the diameter?

10 A helicopter landing circle needs to be painted on an airfield. If the circle needs to be a minimum of 153 m^2, what should the radius be, correct to the nearest metre?

12 Use simple formulae expressed in words

This mathematical technique involves basic algebra. Here you will only be required to use simple algebraic formulae.

EXAMPLE 1

If a book on child care is priced at £7 and a book store sells 23 copies, what is the total revenue from the book?

Total revenue is calculated by:

Total revenue = Price × Quantity

Substitute price and quantity into the formula:

Total revenue	=	Price × Quantity
	=	£7 × 23
	=	£161

The performance criteria analysis for this example appears in Fig 12.1.

	Performance Criteria	Pass or NYA
1	techniques appropriate to the task are selected and used by the individual	Pass. used simple algebra
2	activities required by the techniques are performed to appropriate levels of precision and in the correct sequence	Pass. correct throughout
3	mathematical terms and mathematical aspects of everyday language are used and interpreted with precision	Pass calculated cost
4	use estimation to check calculations	NYP no estimate
5	calculations and results obtained are correct	Pass all correct
6	aids are used correctly	Pass Calculator
7	conclusions / generalisations / predictions drawn from results are valid	NYP Did not double check
8	clear justification is given for the interpretation of results	with $x = ?$ in original problem.

Fig 12.1

The following exercise uses these formulae:

$$\text{Total cost} = \text{Fixed cost} + (\text{Variable cost} \times \text{Quantity})$$

$$\text{Net profit} = \text{Gross profit} - \text{Expenses}$$

$$\text{Stock available} = \text{Opening stock} + \text{Purchases}$$

$$\text{Gas bill} = \text{Fixed charge} + (\text{Cost per unit} \times \text{Number of units used})$$

$$\text{Average cost per mile} = \frac{\text{Cost of petrol}}{\text{Number of miles travelled}}$$

$$\text{Net wage} = \text{Gross wage} - \text{Deductions}$$

Exercise 12.1

1 What is the total cost of producing 500 items, if the fixed cost is £10 000 and the variable cost is 35p?

2 Gross profit for *Head and Sons* is £505 600. If expenses totalled £17 850, what is the net profit?

3 *The Shoe Shop* holds 564 pairs of shoes in stock. If 75 more pairs have recently been purchased, what is the stock available?

4 A hospice used 580 units of gas in a quarter. There is a fixed charge of £13.50 and a cost per unit of 56.8p. Calculate its total gas bill.

5 A salesman travelled 27 500 miles in one year. If the petrol cost for the year was £2350, what is the average cost per mile?

6 A man earns £980 per month. He pays £205 income tax and £55 pension. What is his net wage?

13 Solve simple equations

The simple equations you are required to solve are *linear* equations. A linear equation, when drawn on a graph, will always produce a straight line. There is an important rule to remember when solving linear equations:

> What you do to one side you must do to the other. (For example, if you add 5 to the right-hand side (RHS), you must add 5 to the left-hand side (LHS).)
>
> OR
>
> Change sides = change signs. (For example, + changes to –, × changes to ÷, and vice versa.)

EXAMPLE 1

(a) Solve $5x + 7 = 27$

$5x = 27 - 7$ (Here + 7 changed sides and became – 7.)

$5x = 20$

$x = \frac{20}{5}$ (Here × 5 changed sides and became ÷ 5.)

$x = 4$

(b) A theatre has 150 seats and the revenue for a single performance is £1680. Assuming every seat was sold, what is the cost of each seat?
Let the cost of each seat be x. Therefore:

$$150x = £1680$$

$$x = \frac{£1680}{150} \quad \text{(Here} \times 150 \text{ changed sides and became} \div 150.)$$

$$x = £11.20$$

(c) A painter spends £80.75 and buys 6 small canvases and 5 large canvases. The large canvases cost £3.50 more than the small canvases. What is the cost of a small canvas?
Let the cost of the small canvas be x. Therefore, the cost of the large canvas will be x + £3.50.

$$6x + 5(x + 3.5) = 80.75 \quad \text{(You must first multiply out the brackets.)}$$

$$6x + 5x + 17.5 = 80.75$$

$$11x = 80.75 - 17.5$$

$$11x = 63.25$$

$$x = \frac{63.25}{11}$$

$$x = 5.75$$

$$\text{Answer} = £5.75$$

The performance criteria analysis for this example appears in Fig 13.1.

	Performance Criteria	Pass or NYA
1	techniques appropriate to the task are selected and used by the individual	Pass. Used a simple equation to solve problem
2	activities required by the techniques are performed to appropriate levels of precision and in the correct sequence	Pass. showed working out and are in correct sequence
3	mathematical terms and mathematical aspects of everyday language are used and interpreted with precision	Pass used x as a substitute
4	use estimation to check calculations	NYA no estimate
5	calculations and results obtained are correct	Pass all correct
6	aids are used correctly	Pass Calculator
7	conclusions / generalisations / predictions drawn from results are valid	Pass valid answer
8	clear justification is given for the interpretation of results	NYP did not double check

Fig 13.1

Exercise 13.1

1 Solve the following equations:

(a) $x + 8 = 15$

(b) $3x + 7 = 22$

(c) $120x - 40 = 200$

(d) $75x - 425 = 175$

(e) $\frac{x}{5} = 12$

2 5 chairs cost £549.95. What is the cost of each chair?

3 It takes 12 hours to make four objects. How long does it take to make one object?

4 A salesman sells 350 objects and earns £962.50. How much does he earn per object?

5 A crew rows 24 miles in three hours. How many miles per hour do they row?

6 Workers receive a basic wage of £35 plus £1.27 per item. If a worker makes 158 items in a week, what will be the gross wage?

7 A construction site buys 10 bags of cement and 60 bags of sand. The cement costs three times as much as the sand. If the total bill is £90, what is the cost of a bag of cement and the cost of a bag of sand?

8 A computer software salesman sells a database package and a spreadsheet package. The database package costs £21.50 more than the spreadsheet package. In a particular week he sold 15 database packages and 21 spreadsheet packages with a total revenue of £1632. What is the cost of the spreadsheet package?

9 A community centre spends £500 on 30 chairs. They buy soft chairs and hard chairs. The hard chairs cost £15 each and the soft chairs cost £20 each. How many of the hard chairs do they buy?

10 A sports centre costs its hall out at £42.50 per hour. The hall can be divided into two basketball courts and two badminton courts or one basketball court and six badminton courts or 10 badminton courts. Assuming the costs are proportional to size, what is the hourly cost of (a) a badminton court and (b) a basketball court?

14 Use networks to solve problems

A network is an interconnected group or system, such as a networked computer system, a road or rail network, and a telephone network. In mathematics, the term network refers to the connection of a set of activities. Managers use network analysis to aid in the planning and control of projects.

Using a pre-designed network to solve problems

EXAMPLE 1

Study the town centre road network in Fig 14.1 (on page 96). Some of the roads allow two-way traffic and some allow only one-way traffic. What route must be taken to get from the shopping centre to the market by car?

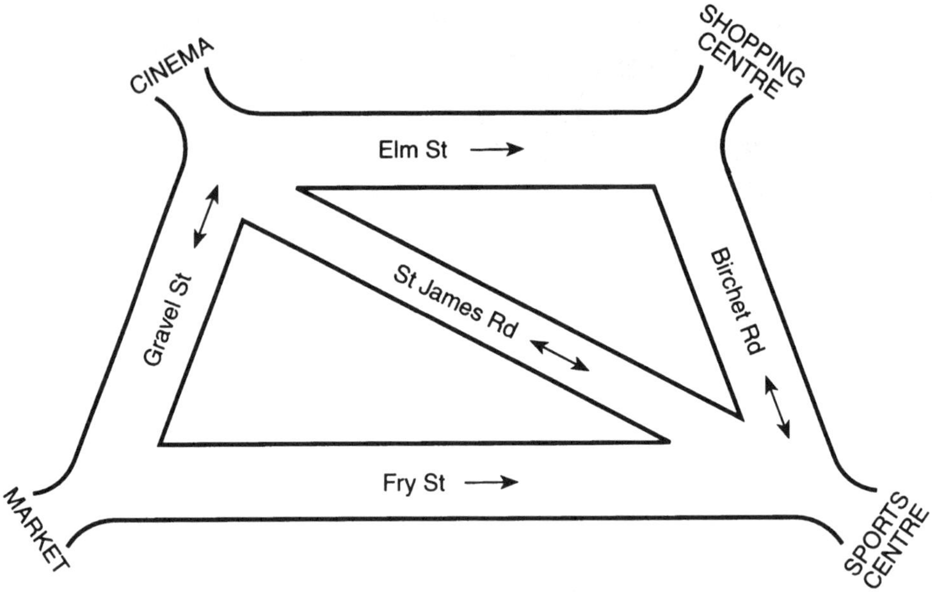

Fig 14.1 Town centre road network

Answer = Along Birchet Road, then along St James Road, then along Gravel Street.

Designing a network to solve problems

When you are designing a network note the following:

- An arrow going from left to right represents an activity.
  ⟶
- The beginning or end of an activity (called the event) is represented by a circle.

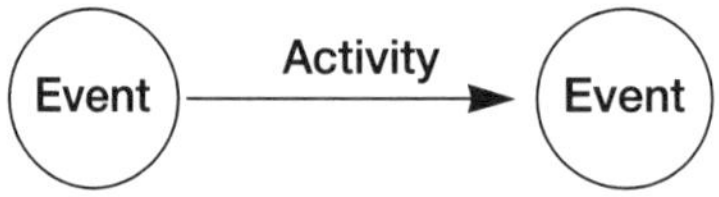

- A complete network should have a start and a finish.

EXAMPLE 2

A student has six activities to complete in a week.

In Step 1 the activities are identified with a letter and displayed in a table (see Table 14.1). (The activities are not in order of priority.)

Table 14.1

Task	*Activity*
A	Get statistical information from library.
B	Analyse statistical information for Assignment 1, Task 1.
C	Attend lecture on how to analyse statistical information.
D	Copy up friend's notes as missed last week's lecture.
E	Write a report for Assignment 1, Task 3 based upon the notes from last week's lecture.
F	Hand in Assignment 1 for marking.

In Step 2, a third column, *Preceding (or prior) activity*, is added to the table. The preceding activity is the activity that comes immediately before the task, e.g. Task C immediately precedes Task B, but not Task F (see Table 14.2).

Table 14.2

Task	*Activity*	*Preceding activity*
A	Get statistical information from library and take to lecture.	—
B	Analyse statistical information for Assignment 1, Task 1.	C
C	Attend lecture on how to analyse statistical information.	A
D	Copy up friend's notes as missed last week's lecture.	—
E	Write a report for Assignment 1, Task 3 based upon the notes from last week's lecture.	D
F	Hand in Assignment 1 for marking.	B, E

A **dash** indicates that there is no preceding activity to this task. This task therefore begins at the start of the network. Fig 14.2 shows the network diagram.

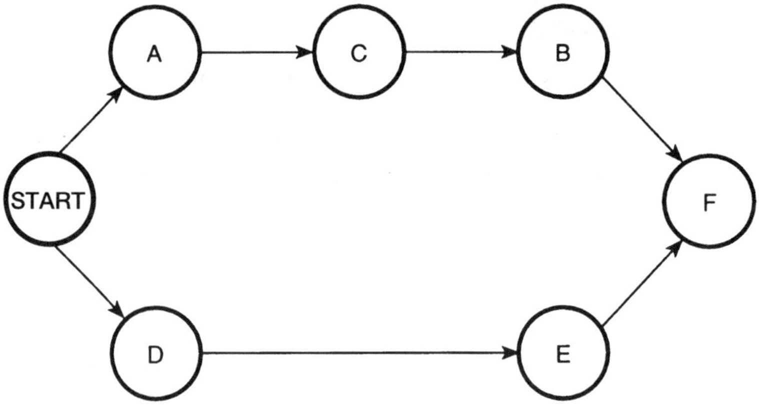

(a) What two activities can be carried out independently?

Answer = The copying up of notes and analysing the statistical information.

(b) What is the final activity?

Answer = The handing in of the assignment.

Exercise 14.1

1 Draw the network diagram for the activity table shown in Table 14.3.

Table 14.3

Activity	*Preceding activity*
A	—
B	A
C	A, B
D	A
E	B
F	C, D, E
G	F

2 Two departments within an organisation are planning to move to a new office building. Table 14.4 identifies the main activities for this move.

Table 14.4

Task	*Activity*	*Preceding Activity*
A	Inform staff about the move.	–
B	Organise removal firm.	A
C	Get packing crates from removal firm.	B
D	Decide on an office plan for new open-plan office.	A
E	Get telephone points installed in new office.	F
F	Get office plan approved by manager.	D
G	Pack crates.	C, E
H	Move office.	G
I	Unpack crates.	H

Draw a network to represent the activities.

3 A manufacturing company is about to produce a new product. The activities involved are shown in Table 14.5.

Table 14.5

Task	*Activity*
A	Collect product specification from R&D.
B	Order raw materials.
C	Order new machinery.
D	A few staff go on a training course to use the new machinery.
E	Raw materials arrive at the warehouse.
F	Raw materials are checked and stored.
G	New machinery arrives at production plant.
H	New machinery is installed.
I	New machinery is tested.
J	The first manufacturing run is started.
K	New product is tested.
L	Finished product is stored in warehouse.

Produce a prior activity table and draw the network of the activities.

15 Carry out calculations in plane and solid shapes

The calculation of area and volume has been covered in previous chapters. In this chapter you are required to take the skills you have learned and use them to calculate the area and volume of shapes that combine different shapes.

EXAMPLE 1

(a) Fig 15.1 shows the dimensions of an athletics track. Calculate the total area.

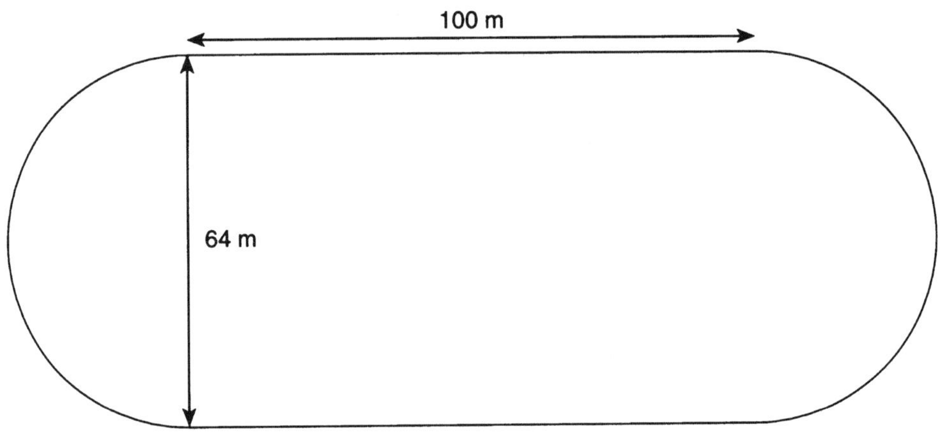

Fig 15.1 Dimensions of an athletics track

The track consists of two two-dimensional (2D) shapes: a rectangle and a circle (two semicircles at either end).

The area of the rectangle is

$64 \text{ m} \times 100 \text{ m} = 6400 \text{ m}^2$

Radius = 32 m

So, the area of the circle is

$3.142 \times 32^2 = 3217.4 \text{ m}^2$

Area of rectangle + area of circle = Total area

$6400 \text{ m}^2 + 3217.4 \text{ m}^2 = 9617.4 \text{ m}^2$

(b) Fig 15.2 shows the dimensions of a swimming pool at a local recreation centre. Calculate the total volume.

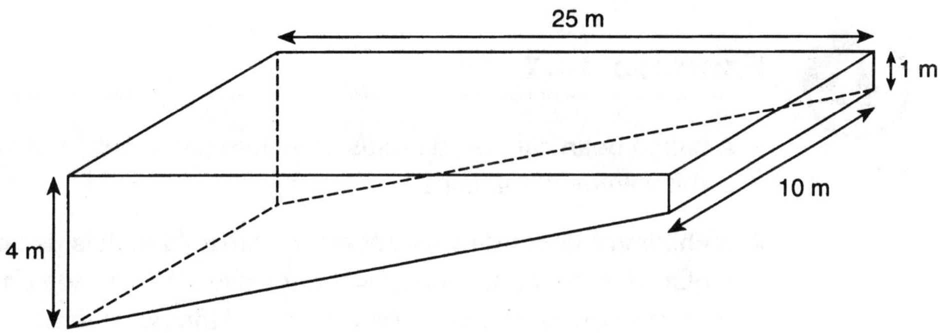

Fig 15.2 Dimensions of a swimming pool

The pool consists of two three-dimensional (3D) shapes: a cuboid and a prism.

The volume of the cuboid is

$1 \text{ m} \times 25 \text{ m} \times 10 \text{ m} = 250 \text{ m}^3$

The volume of the prism is

$\frac{1}{2} \times 3 \text{ m} \times 25 \text{ m} \times 10 \text{ m} = 375 \text{ m}^3$

The volume of the cuboid + the volume of the prism = Total volume.

$250 \text{ m}^3 + 375 \text{ m}^3 = 625 \text{ m}^3$

The volume of a cylinder is the area of the cross-section × height (length).
$\pi r^2 h$

EXAMPLE 2

A cylinder has a radius of 20 cm and a height of 65 cm. What is the volume of the cylinder?

The formula is $\pi r^2 h$.

$3.142 \times 20^2 \times 65 = 81\ 692$

Answer = 81 692 cm^3

Exercise 15.1

1. A baked bean can has a radius of 33 mm and a height of 130 mm. What is the volume of the can?
2. A children's concrete playground is 10 m × 25 m. It is decided to replace the concrete with grass, but leave a 1.5 m wide path all the way round the grass. What will be the area of the pavement?
3. A pipe has an outside diameter of 51 mm and a bore of 48 mm. What is the cross-sectional area of the pipe?
4. What is the volume of a hot-tub that measures 180 cm in diameter and 92 cm in height?
5. A thin sheet of metal is used to construct an open container with a base of 65 cm × 24 cm and a height of 1.2 m. Assuming that the edges do not overlap, calculate the area of the sheet metal used in its construction. Give your answer in both square centimetres and square metres.
6. A piece of wood 1 m × 1.5 m is used to make an occasional table. The table top is circular with a diameter of 55 cm. The support is two rectangles 50 cm × 65 cm slotted together. Calculate the area of wood wasted.
7. A football pitch is 110 m × 75 m. The groundsman uses a roller 1.2 m in diameter and 1.7 m wide to level the pitch. Calculate how many complete revolutions the roller has to make to level the ground.

16 Solve equations or simple inequalities

In a previous chapter you were required to solve simple equations. This section considers more complicated equations such as simultaneous equations. Simultaneous equations are two equations that are connected. The x values in the two equations are identical. The y values in the equations follow the same rule.

Inequalities are equations where one term is greater than or less than the other term.

You need to remember and know the meaning of the following symbols:

$>$ greater than
$\geqslant$ greater than or equal to
$<$ less than
$\leqslant$ less than or equal to

Simultaneous equations and inequalities are needed in industry because it is not always possible to expect things to run smoothly. For example, two machines may be used to manufacture two different products. Only one product can be manufactured at a time on each machine. Simultaneous equations are used to help solve the problem of how many of each product will give the best usage of machinery.

Inequalities are needed because it is not always possible to have a perfect solution. For example, you may have five workers in a factory available for eight hours in a particular day. This does not mean that you have 40 hours of production time. One worker may not be able to use a machine another worker uses. Machines or workers may be idle while waiting for others to finish.

Solving simultaneous equations

EXAMPLE 1

(a) Solve the simultaneous equations:

$$x + 3y = 17 \qquad (1)$$

$$4x - 3y = 8 \qquad (2)$$

Add Equation (1) and Equation (2).

$$(x + 4x) + (3y - 3y) = 17 + 8$$

$$5x + 0 = 25$$

$$5x = 25$$

$$x = \frac{25}{5}$$

$$x = 5$$

Substitute 5 into Equation (1).

$$5 + 3y = 17$$

$$3y = 17 - 5$$

$$3y = 12$$

$$y = \frac{12}{3}$$

$$y = 4$$

Check in Equation (2).

$$(4 \times 5) - (3 \times 4) = 8$$

$$20 - 12 = 8$$

(b) Solve the simultaneous equations:

$$2x + 6y = 20 \qquad (1)$$

$$3x + 2y = 16 \qquad (2)$$

Multiply Equation (1) by 3.

$$6x + 18y = 60 \qquad (3)$$ (This becomes Equation (3).)

Multiply Equation (2) by 2.

$$6x + 4y = 32 \qquad (4)$$ (This becomes Equation (4).)

Subtract Equation (4) from Equation (3).

$$(6x - 6x) + (18y - 4y) = 60 - 32$$

$$0 + 14y = 28$$

$$14y = 28$$

$$y = \frac{28}{14}$$

$$y = 2$$

Substitute 2 into Equation (1).

$$2x + 12 = 20$$

$$2x = 20 - 12$$

$$2x = 8$$

$$x = \frac{8}{2}$$

$$x = 4$$

Check in Equation (2).

$$(3 \times 4) + (2 \times 2) = 16$$

$$12 + 4 = 16$$

Exercise 16.1

Solve the following simultaneous equations for *x* and *y*.

1 $x + 5y = 13$

$6x - 5y = 8$

2 $2x + 6y = 32$

$4x + 3y = 37$

3 $4x + y = 48$

$3x - 3y = 6$

4 $3x + 2y = 79$

$2x + 5y = 115$

5 $7x - 5y = 8$

$2x + 7y = 36$

Solving simultaneous equation problems

EXAMPLE 2

A theatre has 300 seats and charges adults £18 and children £15. At one performance the total revenue was £5160. How many adults and how many children attended the performance?

Let the number of adults be x and the number of children be y.

$$x + y = 300 \quad (1)$$

$$18x + 15y = 5160 \quad (2)$$

Multiply Equation (1) by 15.

$$15x + 15y = 4500 \quad (3)$$

Subtract Equation (3) from Equation (2).

$$
\begin{aligned}
(18x - 15x) + (15y - 15y) &= 5160 - 4500 \\
3x + 0 &= 660 \\
3x &= 660 \\
x &= \frac{660}{3} \\
x &= 220
\end{aligned}
$$

Substitute 220 in Equation (1).

$$
\begin{aligned}
220 + y &= 300 \\
y &= 300 - 220 \\
y &= 80
\end{aligned}
$$

Answer = 220 adults and 80 children were at the performance.

Exercise 16.2

1 A bill for £195 was paid with £5 notes and £10 notes. If 24 notes were used, how many £5 notes were there?

2 Two machines, A and B, are used to make Product *x* and Product *y*. Product *x* needs 1 hour on Machine A and 3 hours on Machine B. Product *y* needs 2 hours on Machine A and 2 hours on Machine B. Machine A can only be used for 16 hours continuously but Machine B can be used for 36 hours. Calculate the number of each product that should be produced to utilise the maximum capacity of the machines.

3 A factory has 500 m of a certain cloth. A dress and a shirt are made out of the cloth. The dress sells at £30 and the shirt sells at £22. If total revenue is £12 264, how many of each item are made?

4 Peter works 38 hours basic and 4 hours overtime and earns £352. Matthew works 40 hours basic and 6 hours overtime and earns £392. Find *x* and *y* if *x* is the basic pay per hour and *y* is the overtime pay per hour.

Solve simple inequalities

Here are some examples of inequalities:

$x > 3$ means an unknown number x is greater than 3 (but not including 3).

$x \geqslant 3$ means an unknown number x is greater than or equal to 3 (includes 3).

The rules for solving inequalities are the same as for solving equations (e.g. change sides = change sign) with one exception: the inequality sign must be reversed when multiplying or dividing both sides by a negative number.

EXAMPLE 3

Solve the inequality:

$$4 - 5x < 20 - 3x$$

Move x values to left-hand side of inequality:

$$-5x + 3x < 20 - 4$$

$$-2x < 16$$

$$x > \frac{16}{-2}$$

$$x > -8$$

Exercise 16.3

Solve the following inequalities:

1 $5 + 4x > 9 + 2x$

2 $11 + 6x \geqslant 18 - x$

3 $22 - 2x > 7 + 3x$

4 $19 + 3x \geqslant 2x - 5$

5 $25 + 2x < 8x - 9$

6 A nurse has two activities to complete in less than 120 minutes. Activity 1 is to roll bandages and Activity 2 is to make up pressure pads. The bandages take two minutes to roll and the pressure pads take five minutes to make. Write an expression for this problem. If she has to have 14 pressure pads, how many bandages can she make?

7 A factory produces widgets (x) and gadgets (y). It takes one hour and 3 kg of raw materials to make a widget. It takes two hours and 1 kg of raw materials to make a gadget. The factory can only produce one item at a time and has only 64 hours per week production time. The factory is also constrained in that they are only delivered 72 kg of raw materials per week. How many of each product should they make to maximise production? (Write an expression for the time constraint and an expression for the raw material constraint and solve the equations.)

17 Use symbolic notation to express the rules of sequences

This mathematical technique is founded upon rules of algebra and as with algebra most sequences are based upon addition, subtraction, multiplication and division.

A sequence is a set of numbers connected by some definite law, for example, the sequence 3, 6, 9, 12, 15, 18, . . . is formed by adding 3 to the previous term.

Science has many formulae that are connected by laws and therefore cover this mathematical technique adequately (kinematics – distance, velocity and acceleration; oscillations and waves; heat flow; DC and AC circuits; radioactive decay).

EXAMPLE 1

A salesman earns a basic pay of £137, plus £4.50 per item sold. How much does he earn when he sells three items?

The sequence is:

£137 + £4.50 + £4.50 + £4.50 = £150.50

This does not satisfy the mathematical technique as we have not used symbolic notation to express the rules of sequences.

The compound interest formula is an ideal example of the use of symbolic notation to express the rules of sequences. Compound interest is a sequence and the formula contains symbolic notation.

EXAMPLE 2

£500 is invested at 7% compound interest for three years. How much is the investment worth after three years?

The sequence is:

	£	
Principal sum invested	500.00	
Plus interest received in year 1	35.00	(7% of £500)
Value of investment at the end of year 1	535.00	
Plus interest received in year 2	37.45	(7% of £535)
Value of investment at the end of year 2	572.45	
Plus interest received in year 3	40.07	(7% of £572.45)
Value of investment at the end of year 3	612.52	

The symbolic notation in the compound interest formula is:

$$A = P\,(1 + r)^n$$

where A = Amount accrued

P = Principal (a sum of money either invested or borrowed)

r = Interest rate (expressed as a decimal)

n = Number of years

To satisfy this mathematical technique you must calculate compound interest with the formula and not through the long, sequence method. So if £500 is invested at 7% compound interest for three years, we have:

$$A = 500\,(1 + 0.07)^3$$

$$A = 500 \times 1.07^3 \qquad (1.07^3 \text{ is } 1.07 \times 1.07 \times 1.07)$$

$$A = 500 \times 1.225043$$

$$A = 612.5215$$

$$A = £612.52$$

To work this out on a scientific calculator you need to find an X^y or Y^x button. Use the following sequence:

(1) (+) (0.07) (=) ___ (X^y) (3) (=) ___ (×) (500) (=) ___

Exercise 17.1

1 £3500 is invested at 9% compound interest. How much is the investment worth after three years?

2 £1500 is borrowed for four years at 12% compound interest. How much has to be paid back at the end of the period?

3 £15 500 is invested at $5\frac{1}{2}$ % compound interest. How much is the investment worth after 12 years?

4 How much interest is to be paid on a loan of £700 for two years at 11% compound interest?

5 How much interest will I get after seven years if I invest £950 at 8% compound interest?

To work out *depreciation* (the reduction in the value of a fixed asset due to wear and tear etc.), we use the compound interest formula with a slight variation. The depreciation formula is:

$$A = P\,(1 - r)^n \text{ (the + has changed to a –)}$$

where A = Value of asset at end of period

P = Original value of asset

r = Depreciation rate (expressed as a decimal)

n = Number of years

Exercise 17.2

1 The value of a machine depreciates at 20% per annum. If it cost £5500 when new, calculate its value after five years.

2 A van cost £15 000 when new. If it depreciates at 12% per annum, what value will it be at the end of five years?

3 A company buys a machine for £30 000 and estimates its life to be ten years. If depreciation is calculated at 14% per annum, what will be the value of the machine at the end of the ten years?

4 A machine that has been depreciated by 8% per annum over five years is now worth £300. What was the original value of the machine?

5 A machine is worth £5000 after five years. If it was depreciated at 16% per annum, what was the original cost of the machine?

Part 3
INTERPRET AND PRESENT MATHEMATICAL DATA

Instant communication is the ideal in any vocation. Data or information can be presented in writing, in drawings, in diagrams and in mathematical expressions. This section introduces you to appropriate methods of presentation for different types of data or information.

Wherever possible, you are required to use the techniques in this part of the book to present the data you have collected in Parts 1 and 2. For example, you can present the information gathered from your survey on a graph, or the data from solving equations can be justified and expanded by using probability.

18 Use mathematical terms to describe common 2D shapes and 3D objects

The mathematical terms you are required to use are shown below:

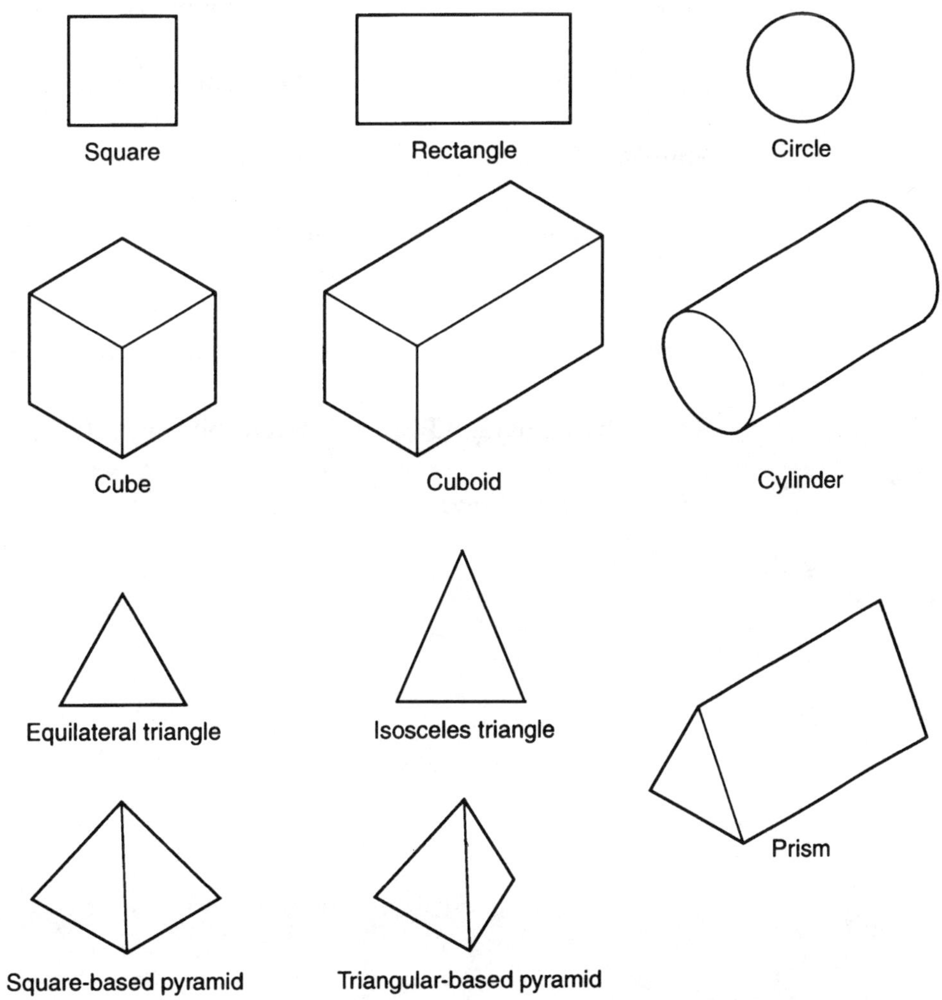

From these mathematical terms you need to be able to identify everyday objects (or objects common to your vocation). For example:

Square	Chess board, table top, computer floppy disk case
Rectangle	Piece of A4 paper, seat on a park bench
Circle	Clock, dart board, coin, satellite dish, compact disk
Cube	Dice, OXO cube, Rubix cube
Cuboid	Filing cabinet, TV, video
Cylinder	Baked bean can, pen,
Equilateral triangle	Give way road sign
Isosceles triangle	Sandwiches you buy from the supermarket in a plastic case
Prism	Roof of a house, Toblerone bar
Square-based pyramid	Church spire, Egyptian pyramids
Triangular-based pyramid	Chocolate sweet

19 Use 2D representation of 3D objects

This mathematical technique requires you to:

- draw and recognise nets of common 3D objects;
- draw 3D objects;
- interpret simple plans and elevations.

Recognising and drawing nets

A net is the two-dimensional representation of a three-dimensional object when opened out, e.g. a cereal box.

Exercise 19.1

1 Name the 3D shapes for the nets in Fig 19.1 (on page 120).

2 Draw the nets for the following objects:

(a) a 5 cm square box
(b) a cassette case 7 cm × 11 cm × 1.5 cm
(c) a Toblerone bar
(d) a baked bean can
(e) a triangular-based pyramid

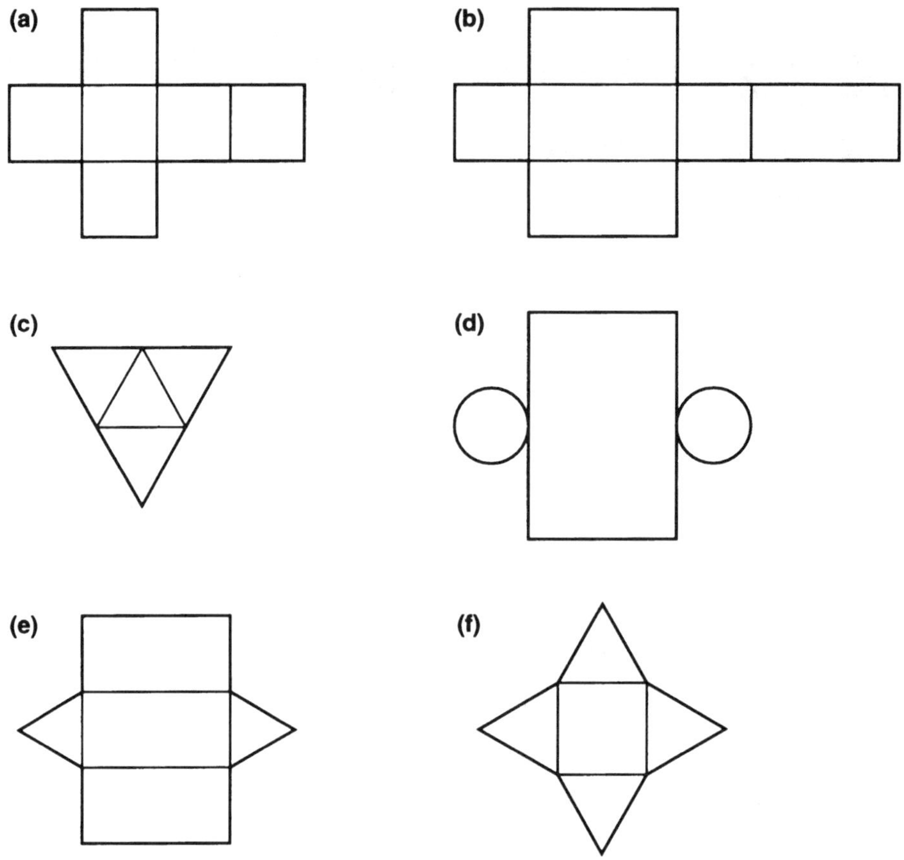

Fig 19.1 Nets for various 3D shapes

Drawing 3D objects

Begin by drawing some of the shapes from Chapter 18. It can be difficult to draw three-dimensional shapes on paper, so isometric grid paper has been provided (see Fig 19.2). Place a plain piece of paper over the isometric grid and use the lines as a guide.

Fig 19.2 Isometric grid

Exercise 19.2

1 Identify the objects in 19.3.

Fig 19.3

2 Draw the following objects:

(a) a filing cabinet with three drawers
(b) an oil drum
(c) a house

Interpret simple plans and elevations

Figure 19.4 is a plan of a kitchen. Learn the symbols for windows and doors (especially the way they open).

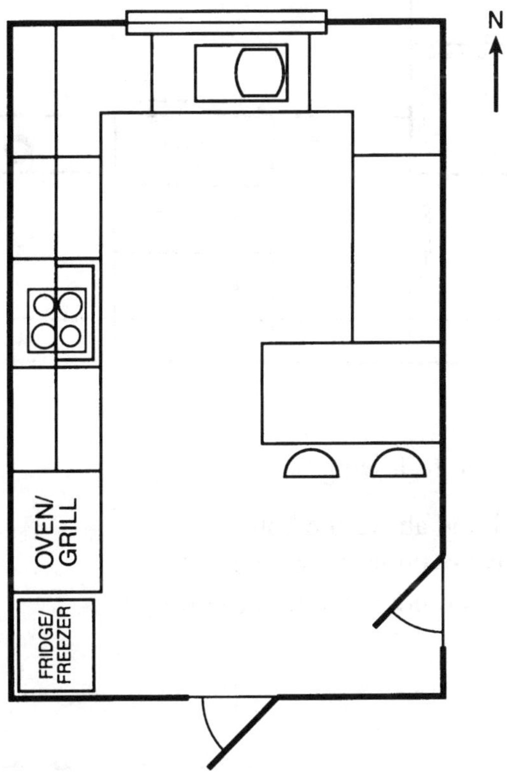

Fig 19.4

The plan shows the following:

- The hob is half way down the room on the west wall.
- The window is on the north wall.
- There is a door on the south wall and at the bottom of the east wall.

- There is a breakfast bar on the east wall next to the door.
- The sink is under the window.

Figure 19.5 shows an elevation of part of the west wall in the kitchen.

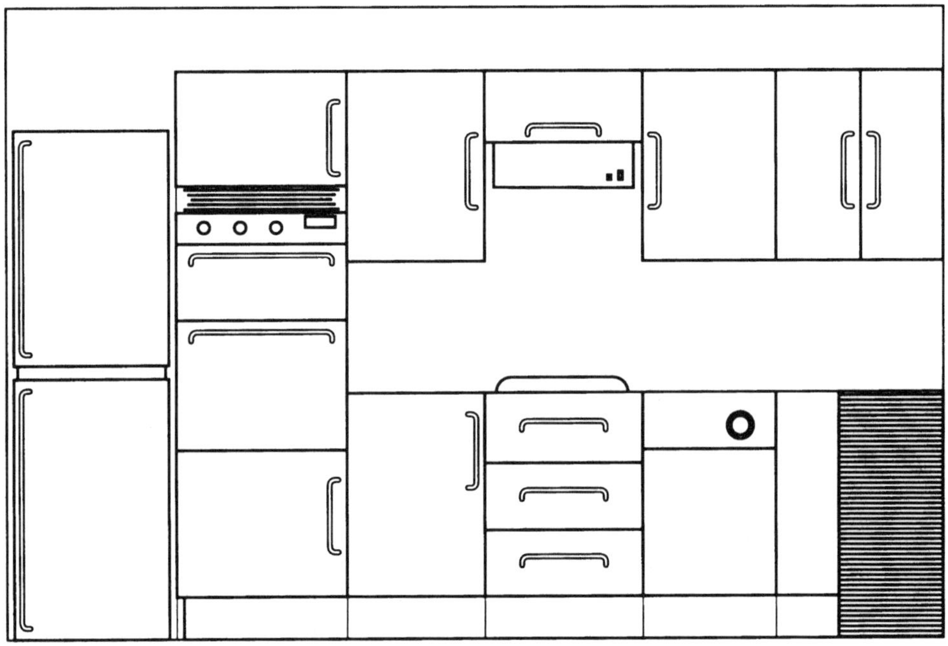

Fig 19.5

The elevation shows the following:

- There is a cooker hood above the hob.
- There is a split-level oven in the kitchen.
- The wall cupboards do not reach to the ceiling.

Exercise 19.3

Study the house plans in Fig 19.6 and answer the questions on page 126.

Ex 19.3 Study house plans in Fig 19.6 & answer the questions which follow.

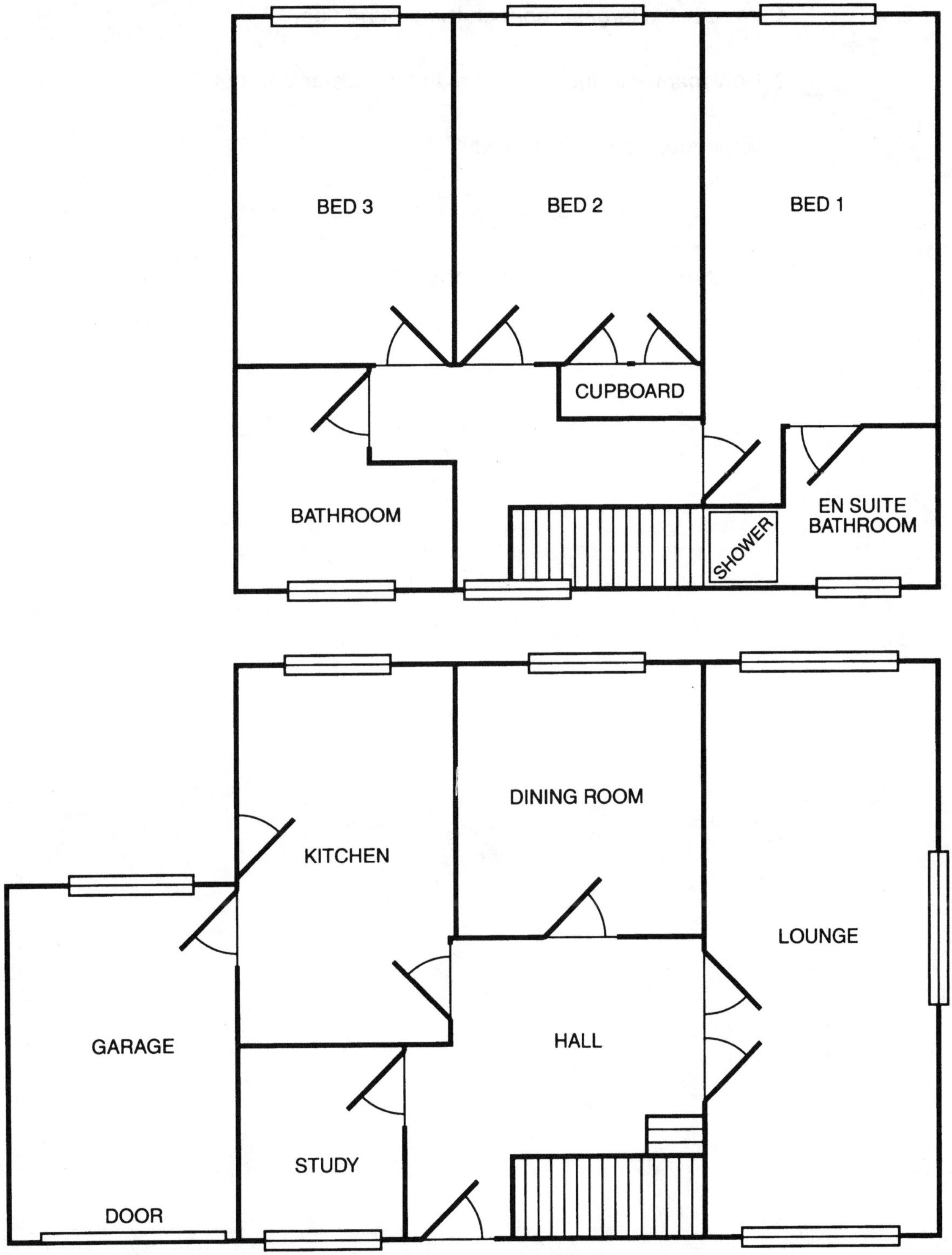

Fig 19.6

1 What is the largest room on the ground floor?

2 How many windows are there in the master bedroom?

3 What room is over the study?

4 Where does the internal door to the garage go from?

5 What room has a built-in cupboard?

20 Construct and interpret statistical diagrams

Statistical diagrams are essentially concerned with displaying data to the greatest advantage. A sales team would be keen to have management and customers see its impressive increase in sales. A line graph, like the one in Fig 20.1, with a broad line indicating sales figures will give an immediately favourable representation of this information.

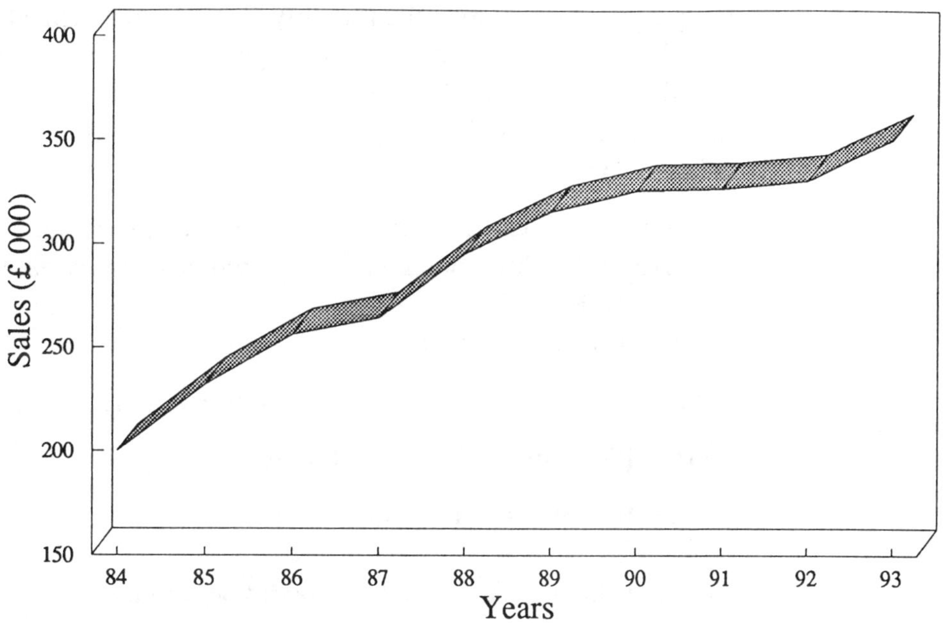

Fig 20.1 Sales for company X

It is necessary to point out that data is either discrete or continuous. You must know the difference. *Discrete data* is any data that has exact values.

For example:

- the number of children in a family
- the number of light bulbs made in a factory
- the number of goals scored in a football match
- the shoe sizes of members of a health club.

Continuous data is any data that can only be measured to a certain degree of accuracy. It cannot assume any exact value. For example:

- the heights of children in a family correct to the nearest cm
- the life of 20 light bulbs to the nearest minute
- the time taken to make a widget to the nearest minute
- the speed of cars passing a checkpoint.

Reminders on construction of graphs

1 The information on the y axis always depends upon the information on the x axis. For example, the monthly sales figures depend upon the month of the year. Therefore, sales figures are on the y axis and months on the x axis. The population of major British cities depends upon the city. Therefore, population is on the y axis and the cities are on the x axis.

y axis	vertical axis	dependent variable
x axis	horizontal axis	independent variable

2 Choose a scale for each axis. Some common scales are:

1 cm = 1 unit   1 cm = 5 units   1 cm = 10 units   1 cm = 25 units

1 cm = 100 units   1 cm = 250 units

1 cm = 1000 units   1 cm = 2000 units

3 When displaying data on a line graph you join the points with a ruler. You only join points freehand if you are using a curve to calculate gradient, etc.

4 You do not join any of the points on a scatter graph.

5 Give a main title to the graph and titles to the x and y axes.

EXAMPLE 1

Figure 20.2 shows the differences between a line graph and a bar chart. Both graphs are displaying the same information.

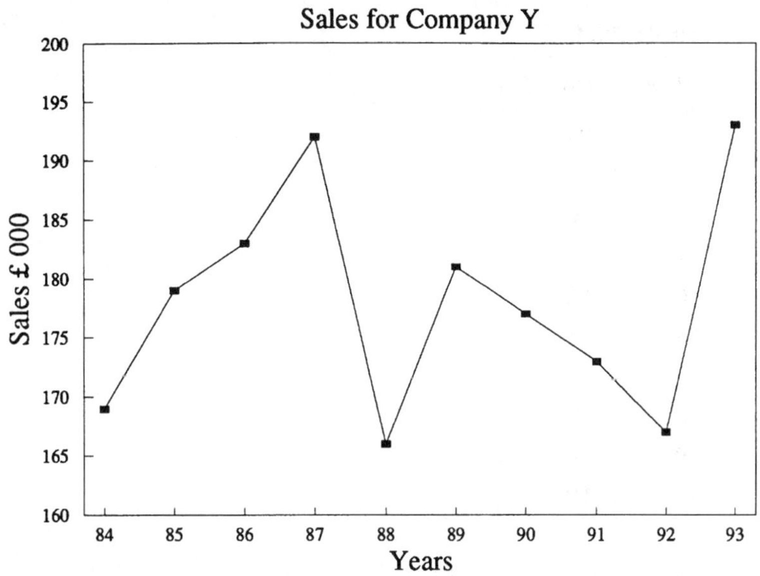

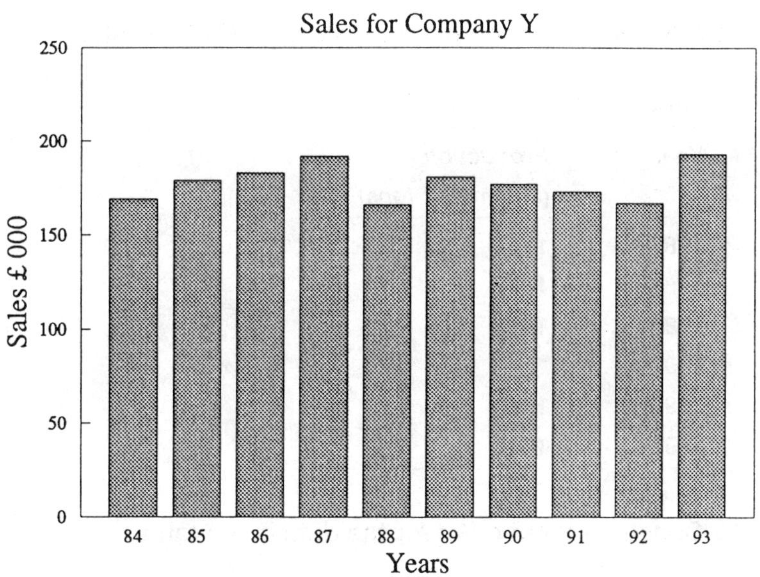

Fig 20.2

Exercise 20.1

1 The number of orders received by a manufacturing company over ten weeks was as shown in Table 20.1.

Table 20.1

Week	*No. of orders*
1	20
2	44
3	56
4	82
5	120
6	110
7	96
8	68
9	50
10	12

Construct a line graph for the data.

2 B & D's production of crude steel, 1987–92, is shown in Table 20.2.

Table 20.2

Year	*Production (000 metric tons)*
1987	174
1988	189
1989	187
1990	178
1991	164
1992	164

Construct a bar chart for the data in the table.

3 The monthly sales figures for Product A and Product B in 199– are shown in Table 20.3.

Table 20.3

Month	*Sales of A (£)*	*Sales of B (£)*
January	2000	1000
February	2800	1500
March	3040	2500
April	3560	3440
May	4600	5020
June	5080	7800
July	5700	8500
August	6800	6000
September	7580	5400
October	7600	4500
November	7800	3000
December	8000	2200

Construct both a bar chart and a line graph for the data in the table.

Scatter graphs

A scatter graph is used to compare or correlate two variables. The main problem with constructing scatter graphs is deciding which variable goes on which axis. Remember the rule:

y axis	vertical axis	dependent variable
x axis	horizontal axis	independent variable

EXAMPLE 2

Figure 20.3 shows three scatter graphs.

In Fig 20.3(a) there is no correlation between marketing expenditure and sales. This means that any increase in marketing expenditure has not necessarily had a positive effect on sales.

In Fig 20.3(b) there is a positive correlation between ice-cream sales and temperature. This means that as the temperature increased, the sales of ice-cream increased.

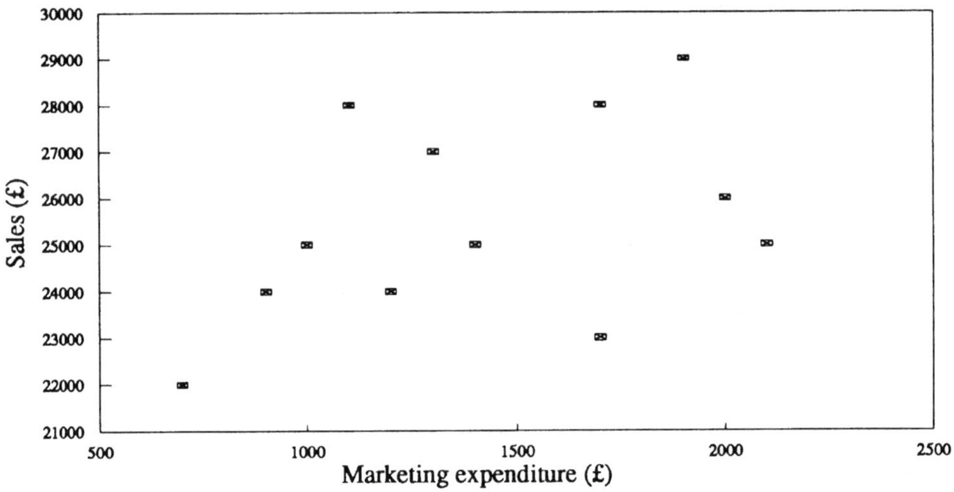

Fig 20.3 (a) Correlation between marketing expenditure and sales

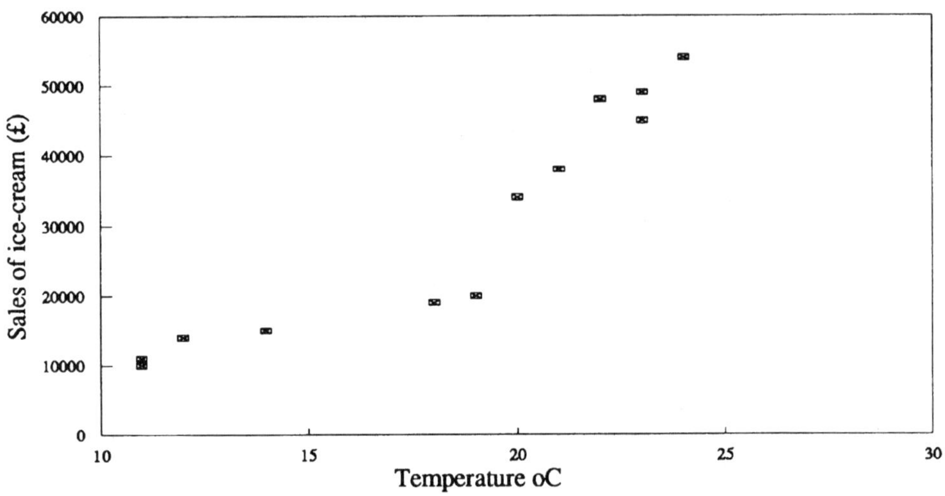

Fig 20.3 (b) Correlation between ice-cream sales and temperature

In Fig 20.3(c) there is a negative correlation between demand and price. This means that there is less demand for expensive items and greater demand for cheaper items.

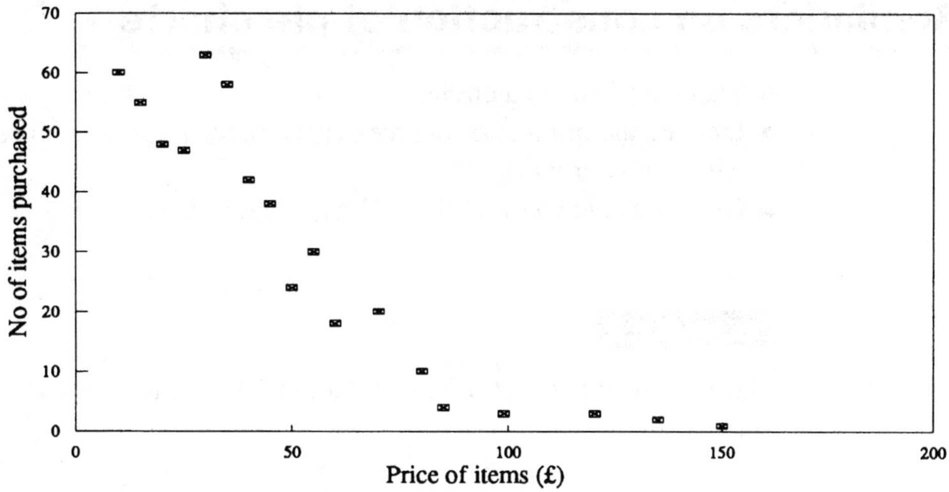

Fig 20.3 (c) Correlation between demand and price

Exercise 20.2

Construct a scatter graph of the information presented in Table 20.4.

Table 20.4

Number of hours of personal study	*Grade for assignments*
13	A
8	C
3	E
7	B
4	D
6	D
8	C
9	D
11	B
10	B
12	A

A is the higher grade and E is the lower grade.

Reminders on construction of pie charts

- There are 360° in a circle.
- Indicate the number of degrees or the percentage of each section somewhere in your work.
- Give your pie chart a title, and name each slice.

EXAMPLE 3

The advertising costs for Brits & Co. in 199– were as shown in Table 20.5.

Table 20.5

Advertising method	*Cost (£000)*
Exhibitions	70
Newspapers	65
Posters	25
Radio	45
TV	95

Add up all the costs.

Total = 300

The number of degrees for each 'slice of pie' are:

Exhibitions $\frac{70}{300} \times 360 = 84°$

Newspapers $\frac{65}{300} \times 360 = 78°$

Posters $\frac{25}{300} \times 360 = 30°$

Radio $\frac{45}{300} \times 360 = 54°$

TV $\frac{95}{300} \times 360 = 114°$

The right-hand column should always add up to 360°. The data is presented in the pie chart in Fig 20.4. (The degrees have been worked out as percentages as well, for ease of interpretation.)

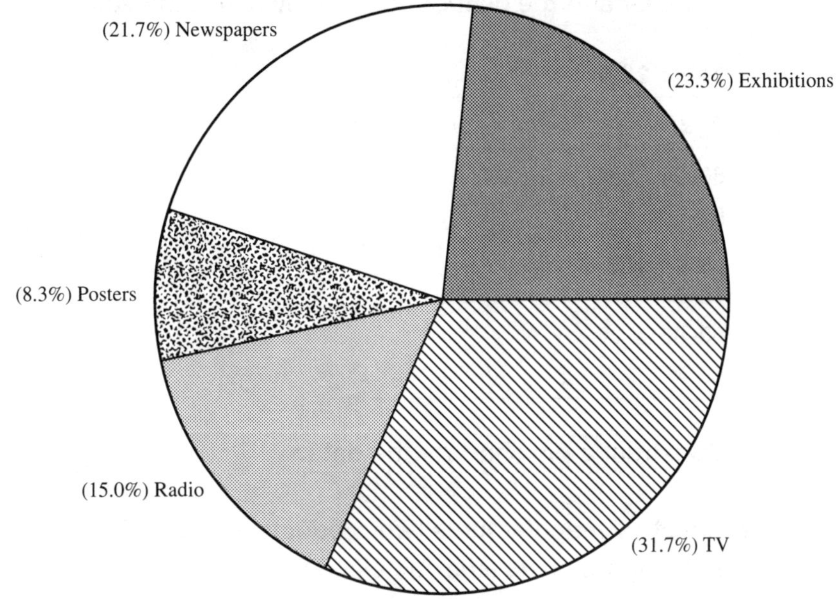

Fig 20.4 Advertising costs of Brit & Co. 199–

The performance criteria analysis for this example appears in Fig 20.5.

	Performance Criteria	Pass or NYA
1	trends / significant features of data are identified correctly	Pass. Construction of a pie chart to show features
2	conclusions / generalisations / predictions drawn from the data are valid	NYP no conclusions drawn
3	mathematical terms and mathematical aspects of everyday language are used and interpreted with precision	Pass. Correct process for construction of a pie chart
4	clear justification is given for the interpretation of data	NYP. No interpretation of data
5	conventions for presenting data are followed accurately	Pass. Correct construction of a pie chart
6	data are represented accurately and to appropriate levels of precision	Pass. Used a valid method of presentation

Fig 20.5

Exercise 20.3

1 The Wendel Nursing Home has 50 residents. Their ages are grouped together and the distribution is shown in Table 20.6.

Table 20.6

Age	*Number of residents*
60–64	5
65–69	9
70–74	13
75–79	15
80–85	6
85 +	2

Construct a pie chart for the data.

2 A factory manufactures two products X and Y. Construct two pie charts to compare costs (see Table 20.7).

Table 20.7

Cost	*Product X (£000)*	*Product Y (£000)*
Direct labour	58	48
Direct materials	30	65
Overheads	27	12
Administration	5	15

Interpret statistical diagrams

Statistical diagrams contain most of their information in the body of the graph or chart. The main title and axes titles also contain information. When you are asked to interpret statistical diagrams write your answer in full (usually one or two sentences).

Use the diagrams on pages 137, 138 and 139 to answer the questions in Exercise 20.4 on page 140.

Fig 20.6 (a)

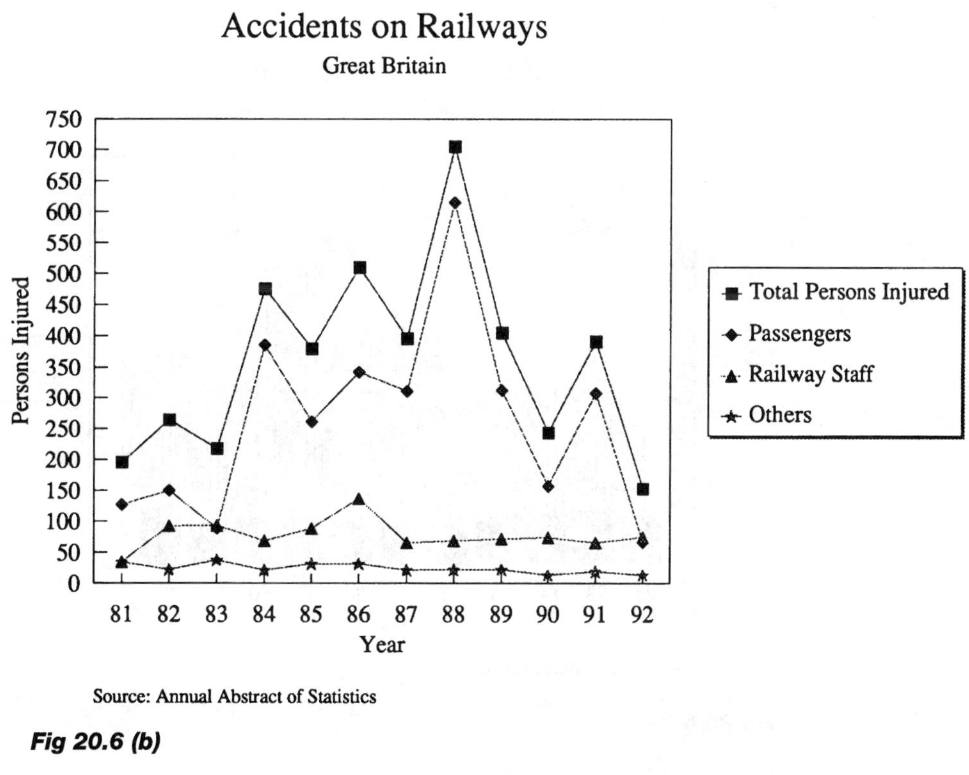

Fig 20.6 (b)

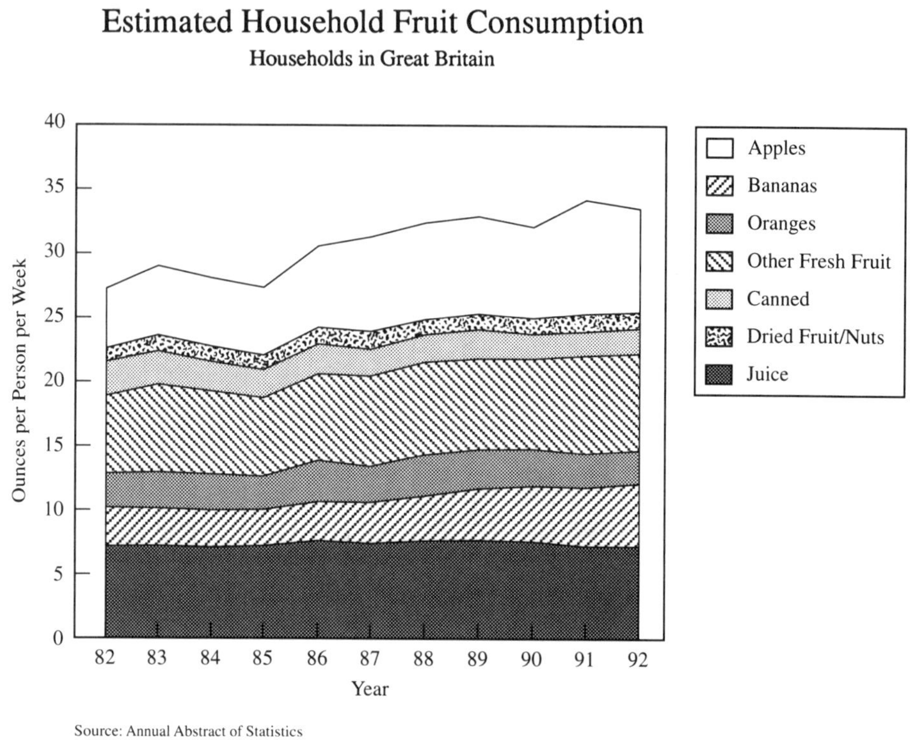

Fig 20.6 (c)

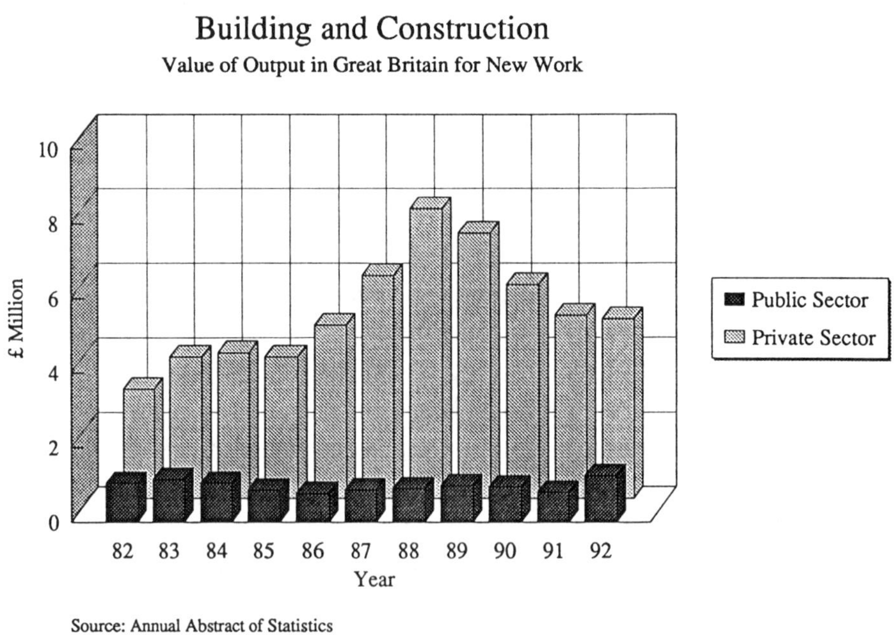

Fig 20.6 (d)

Injuries at Work
Major 1991

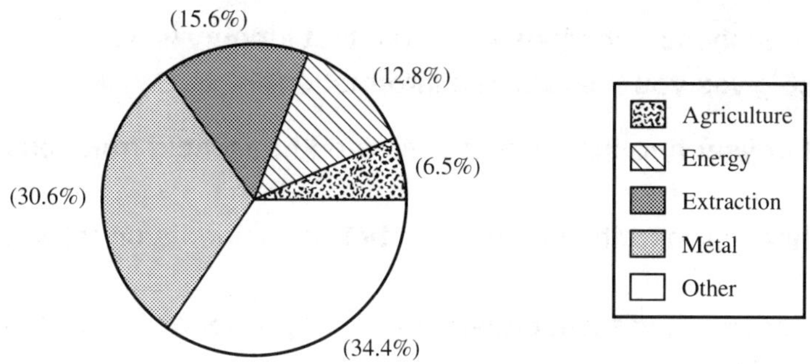

Source: Annual Abstract of Statistics

Injuries at Work
Over 3 days 1991*

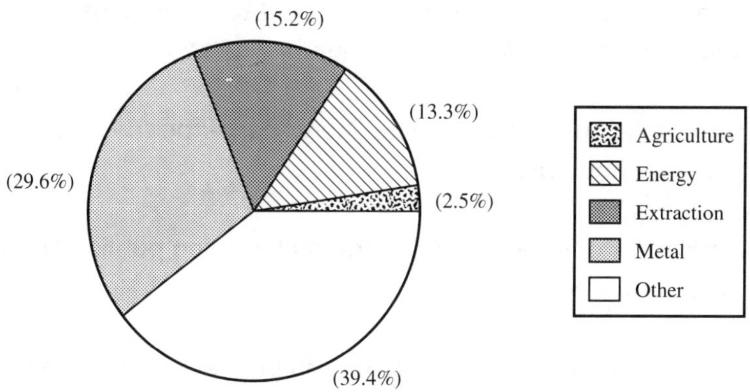

Source: Annual Abstract of Statistics

* Injuries causing incapacity for normal work for more than 3 days.

Injuries at Work
Fatal 1991

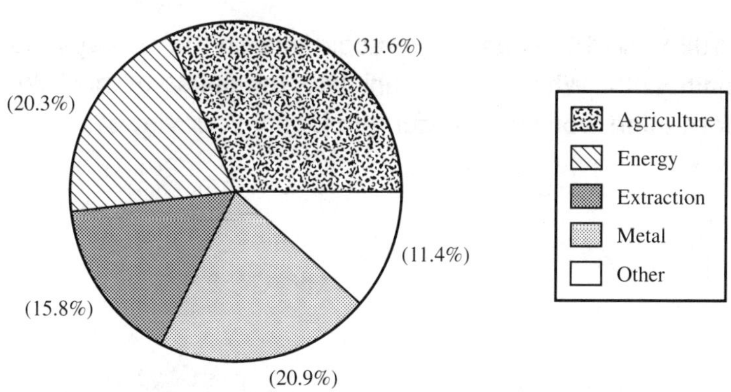

Source: Annual Abstract of Statistics

Fig 20.6 (e)

Exercise 20.4

To answer these questions you must first identify which of the diagrams in Fig 20.6 gives you the required information.

1 What year had the highest estimated household fruit consumption?

2 What year had the highest number of accidents on railways?

3 What year had the highest number of passenger movements by sea and air?

4 What is the highest value of output in new constructions?

5 Which industry had the greatest percentage of injuries at work in 1991 (excluding 'other')? What type of injury was it?

6 What fruit do the British people like to consume the most? What fruit is the least popular?

7 Describe the differences in the trends between public sector and private sector construction.

8 Describe the differences in the trends between sea and air passenger movements.

9 What category of traveller suffered the greatest number of accidents on the railways?

10 In 1984 and 1988 passenger accidents on the railways rose significantly. What do you think was the cause of this? What extra information would you require to justify your views?

21 Use symbols and diagrams

Information can be communicated in a variety of ways.

- Write it down, e.g. in a report or memo.
- Display it, e.g. graphs or charts.
- Draw it, e.g. diagrams or symbols.

The purpose of using diagrams and symbols is to get a message across in the easiest possible way. Dangers are warned of with symbols, e.g. road signs for dangerous bends or falling rocks. Figure 21.1 shows some everyday symbols.

Fig 21.1

Some useful diagrams are shown in Fig 21.2.

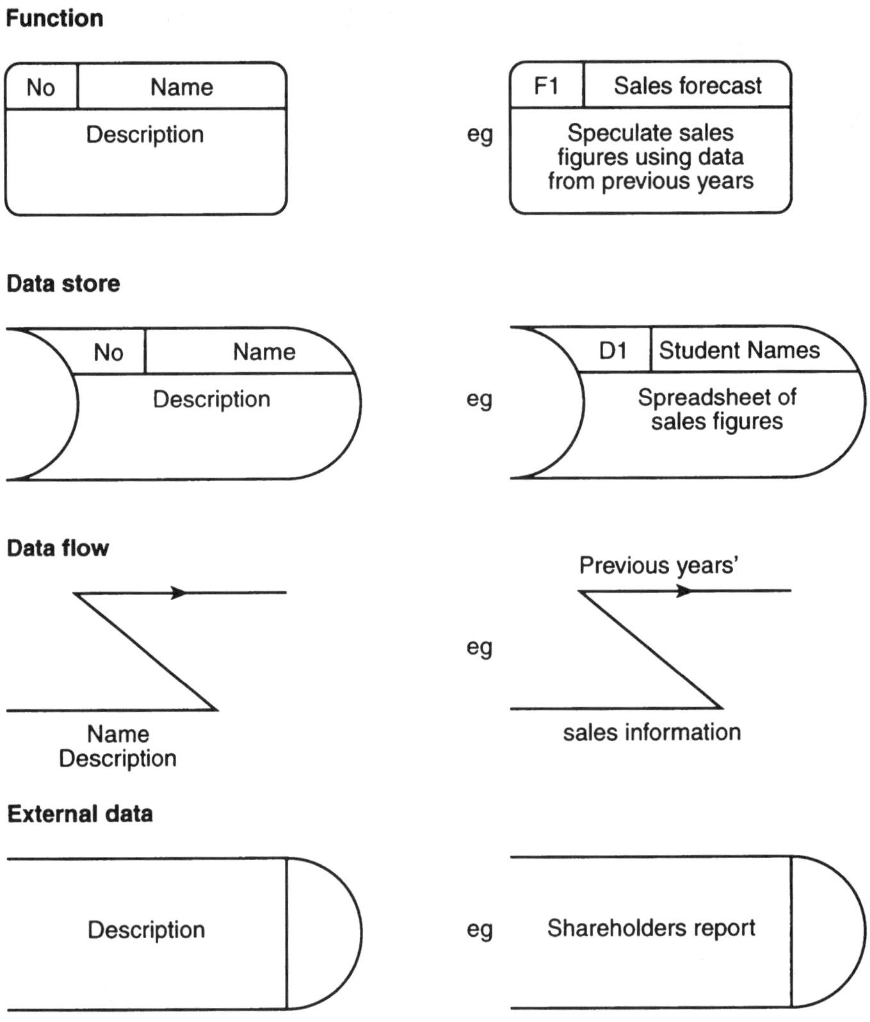

Fig 21.2

Some common Greek symbols used in mathematics appear in Table 21.1.

Table 21.1

Symbol	*Name*	*Common uses*
α	alpha	Statistics
β	beta	
γ	gamma	
δ	delta	Differential equations
Δ	capital delta	Change
ε	epsilon	
θ	theta	Angles
λ	lambda	Linear variable
μ	mu	
ξ	xi	
π	pi	Constant 3.142
ρ	rho	
σ	sigma	
Σ	capital sigma	Summation
ϕ	phi	
χ	chi	
φ	psi	
ω	omega	Change of an angle
Ω	capital omega	Ohms

22 Use the mean and range of a set of data

The *mean* is the average of a set of data. It is found by adding all the observations together and dividing by the number of observations.

The *range* is the difference between the smallest value and the largest value of a set of data. It is found by subtracting the smallest value from the largest value.

EXAMPLE 1

(a) Ten light bulbs were tested to see for how long they would burn. The results, in hours, are:

200 203 214 198 212 198 220 199 199 201

Find the mean and the range.
The mean is:

$$\frac{200 + 203 + 214 + 198 + 212 + 198 + 220 + 199 + 199 + 201}{10} = \frac{2040}{10} = 204$$

The range is:

$220 - 198 = 22$

Answer
Mean = 204 hours
Range = 22 hours

By calculating both the mean and the range we can decide how close to the original data the average is. If the range is large, we know that somewhere in the set of data is an extreme value. In this example, the extreme value is 220 hours.

(b) Compare the performance of two salesmen by using the mean and the range of their weekly orders (see Table 22.1).

Table 22.1

Week	1	2	3	4	5	6
Salesman A	33	86	50	91	56	98
Salesman B	74	59	80	52	71	48

Salesman A Mean = 69 Range = 65 Salesman B Mean = 64 Range = 32

Salesman A has a higher mean (69 orders) than Salesman B (64 orders). However, B has shown more consistency in his work by having a smaller range.

Fig 22.1 shows the performance criteria analysis for this example.

	Performance Criteria	Pass or NYA
1	trends / significant features of data are identified correctly	Pass. Identified, mean, range and extreme value
2	conclusions / generalisations / predictions drawn from the data are valid	Pass. generalisations valid as have mathematical evidence
3	mathematical terms and mathematical aspects of everyday language are used and interpreted with precision	Pass. Calculated mean and range correctly
4	clear justification is given for the interpretation of data	NYP. no interpretation of data
5	conventions for presenting data are followed accurately	Pass. In words and numbers
6	data are represented accurately and to appropriate levels of precision	Pass All calculations correct.

Fig 22.1

Exercise 22.1

1 Calculate the mean and range for the following:

(a) 10, 9, 11, 8, 13, 7, 14, 16

(b) 114, 87, 131, 75, 151, 66, 173, 57, 195, 50, 44, 57

(c) 98, 95, 92, 94, 91, 99, 96

(d) 251, 281, 225, 249, 258, 272, 256, 262, 258, 268

2 Calculate the mean and the range for the monthly sales figures for Product A and Product B in 199–, as shown in Table 22.2.

Table 22.2

Month	*Sales of A (£)*	*Sales of B (£)*
January	2000	1000
February	2800	1500
March	3040	2500
April	3560	3440
May	4600	5020
June	5080	7800
July	5700	8500
August	6800	6000
September	7580	5400
October	7600	4500
November	7800	3000
December	8000	2200

If the company were to consider ceasing to produce one of the products, which would you recommend and why?

3 Compare the performance of two machines that pack sugar. Ten sample packets are weighed, correct to the nearest gram, and their weights recorded below. Calculate the mean and the range for each machine. What recommendations would you make to the company regarding the two machines?

Machine A	495, 496, 496, 498, 499, 500, 500, 501, 501, 504
Machine B	492, 494, 494, 499, 500, 501, 503, 504, 506, 507

23 Identify all the outcomes of combining two independent events

This mathematical technique is founded on basic probability. Events have different probabilities from 0 to 1.

- A probability of 0 means the event cannot happen (e.g. flying unaided).
- A probability of 1 means the event will happen (e.g. dying).

Independent events are events that are not affected by each other. (The toss of a coin is not affected by another toss of a coin.)

The probability of an event occurring is defined as

$$P\text{ (the event)} = \frac{\text{total number of favourable outcomes}}{\text{total number of possible outcomes}}$$

EXAMPLE 1

(a) What is the probability of a thrown die landing on 4?

$$P\text{ (die landing on 4)} = \frac{1}{6} \quad \frac{\text{total number of favourable outcomes}}{\text{total number of possible outcomes}}$$

This is an independent event

(b) If a bag contains three blue socks, three green socks and four red socks, what is the probability of picking out a blue sock?

$$P\text{ (blue sock)} = \frac{3}{10} = 0.3 \quad \frac{\text{total number of favourable outcomes}}{\text{total number of possible outcomes}}$$

(c) What is the probability of picking a blue *or* red sock out of the bag described in (b)?

P (blue sock) = 0.3

P (red sock) = $\frac{4}{10}$ = 0.4

P (blue *or* red sock) = 0.3 + 0.4 = 0.7

These are two independent events. (Here we *added* the two probabilities together.)

(d) If two socks are taken out of the bag described above, what is the probability of picking out a blue sock *and* a red sock?

P (blue sock) = 0.3

P (red sock) = 0.4

P (blue *and* red sock)= $0.3 \times 0.4 = 0.12$

This is combining two independent events. (Here we *multiplied* the two probabilities together.)

Figure 23.1 shows the performance criteria analysis for this example.

	Performance Criteria	Pass or NYA
1	trends / significant features of data are identified correctly	Pass. Probabilities of independent events identified
2	conclusions / generalisations / predictions drawn from the data are valid	Pass valid because correct answer
3	mathematical terms and mathematical aspects of everyday language are used and interpreted with precision	Pass. Used p(event) correctly
4	clear justification is given for the interpretation of data	Pass. Stated type of event and whether add or multiply
5	conventions for presenting data are followed accurately	Pass. presented as a decimal or fraction
6	data are represented accurately and to appropriate levels of precision	Pass Correct answer

Fig 23.1

Exercise 23.1

1 What is the probability that a tossed coin will land on tails?

2 What is the probability that a rolled die will show:

(a) a three
(b) an odd number
(c) a score more than 4?

3 A box contains 12 tennis balls and 13 cricket balls. What is the probability of selecting:

(a) a cricket ball
(b) a cricket ball and a tennis ball?

4 Cakes on a supermarket shelf were inspected for their sell-by dates. Ten cakes had five days before their sell-by date expired, six cakes had three days and four cakes had one day. What is the probability that:

(a) a cake with only one day before the sell-by date expires is chosen;
(b) two cakes chosen will both have five days before the sell-by date expires?

5 A company analysed the number of days in a row employees were late for work. Table 23.1 shows the results.

Table 23.1

Number of consecutive days late for work	1	2	3	4	5	6	7
Number of employees	12	10	8	5	3	1	1

What is the probability that:

(a) an employee is late for work on only one day;
(b) an employee is late for work three days in a row;
(c) an employee will be late for work four days or more in a row;
(d) two employees will be late for work two days in a row?

24 Given the probability of exclusive events, calculate the probability of a combined event

This chapter expands upon Chapter 23. The mathematics is the same, but you are required to have a greater understanding of probability.

Probability is defined as putting uncertainty into numbers. It is used in industry and business to help solve problems. For example, industry uses probability:

- to determine whether a production line is turning out products at a fast enough rate;
- to determine whether accounts will be paid on time or goods delivered on time;
- to determine whether a product on a production line will be faulty when finished;
- to determine how many letters in a company's internal mailing system get lost.

The probability of an event occurring

The probability of an event occurring is based upon experiment. The probability that a coin will land on tails half the number of times it is tossed has been tested by experiment. The probability that a letter will get lost in a company's internal mailing system is also based upon experiment. In this case the experiment is determined by any records kept of people losing letters.

The probability of an event occurring is not always fixed. For example, the probability of a train being a steam train was one in three in 1960. In

1994 it is no longer the case. The setting of insurance premiums has its base in probability experiment and the probabilities regularly change. In the early 1990s insurance premiums rose considerably. This was due to increases in crime, e.g. crimes such as car thefts and household burglaries. As the probability of a crime occurring increased, the probability that an insurance company would have to pay out insurance also increased. This resulted, therefore, in premiums being raised.

Estimating the probability of an event

Recent, valid data is needed to estimate the probability of an event occurring. This is usually collected by observation. When the information is collected, it can be entered into a formula to calculate the probability. The formula is:

$$P\,(\text{E}) = \frac{\text{the number of times the event occurred}}{\text{the number of times the experiment was performed}}$$

EXAMPLE 1

A production manager wishes to know the probability of an item being faulty when made. He checks 850 consecutive items for faults and records 34 faulty items. What is the probability of an item being faulty?

$$P\,(\text{E}) = \frac{\text{the number of times the event occurred}}{\text{the number of times the experiment was performed}}$$

$$P\,(\text{faulty item}) = \frac{34}{850} = \frac{1}{25}$$

The production manager now knows that 1 in 25 items will be faulty. This will help when filling orders. If an order is placed for 400 items the manager knows that 416 items must be made.

Exercise 24.1

1 An organisation's internal mailing system lost two letters out of 200. What is the probability of a letter going missing?

2 A courier company delivered 54 packages out of 450 late. What is the probability that a delivery will be late?

3 Frisk and Co. had 946 customers out of 1000 pay their account on time. What is the probability that an account will not be paid on time?

4 Tannica and Sons make two products. A production time of 30 minutes is set for Product A and 35 minutes for Product B. The time taken to produce 500 of Product A and 428 of Product B is recorded and analysed. 349 of Product A were produced within the 30 minutes. 296 of Product B were produced within the 35 minutes. Which production run is more efficient, Product A or Product B?

5 A component is made for a speedboat engine. Forty engines are fitted with the component, but two fail to work. What is the probability that an engine fitted with the component will not work?

In Exercise 24.1 you were given the information to calculate probability. In Chapter 4 you were required to design and use an observation sheet. The information you collect from this could enable you to calculate probability. Perhaps your observation could be to test goods on a production line, for example.

Answers to exercises

EXERCISE 1.2

1 (a) 500 cm
(b) 3 cm
(c) 27 cm
(d) 427 cm

2 (a) 210 mm
(b) 3150 mm
(c) 46 mm
(d) 5000 mm

3 (a) 9 ft
(b) 4 ft
(c) 8 ft
(d) $2\frac{1}{2}$ ft

4 (a) 2 yd
(b) 4 yd
(c) 3520 yd
(d) 3 yd

5 (a) 5 m
(b) 3000 m
(c) 18 m
(d) 4870 m

6 (a) 17 km
(b) 5 km
(c) 6.4 km
(d) 64.3 km

7 (a) 2 g
(b) 4000 g
(c) 8.3 g
(d) 17 540 g

8 (a) 7 kg
(b) 25 kg
(c) 3000 kg
(d) 0.67 kg

9 (a) 32 oz
(b) 672 oz
(c) 71 oz
(d) 352 oz

10 (a) 4lb
(b) 70 lb
(c) 3 lb
(d) 30 lb

11 (a) 2000 ml
(b) 300 ml
(c) 1450 ml
(d) 740 ml

12 (a) 5l
(b) 2l
(c) 3.56l
(d) 4.9l

13 (a) 2 pt
(b) 64 pt
(c) $2\frac{1}{2}$ pt
(d) 33 pt

14 (a) 2 gal
(b) 3 gal
(c) $5\frac{1}{2}$ gal
(d) $4\frac{1}{2}$ gal

EXERCISE 1.3

1 (a) 3 hr 50 min
(b) 2 min
(c) 19 hr 31 min
(d) 3 days 2 hr

2 (a) 2 hr 30 min
(b) 2 days 12 hr
(c) 30 sec
(d) 1hr 18 min 17 sec

EXERCISE 2.1

1 (a) 2.44 m
(b) 8.82 lb
(c) 1.1 gal

2 154.32 lb

3 11.44 gal

4 101.6 mm

5 14.48 km

EXERCISE 2.2

1 45.46 l

2 77.23 km

3 16.12 hectares

4 17.64 oz

5 23.62 in

EXERCISE 2.3

1 70 mph

2 1.7 kg/cm^2

3 35°C

4 54 l

5 112 km

EXERCISE 2.4

1 39

2 55 yd^2

3 9°C

4 Car B

5 Carpet A is cheaper.

6 – 14°F

EXERCISE 4.1

1 You should have produced a tally chart and frequency distribution which looks like Fig A.1.

Length of Stay	Tally	Frequency
15 minutes	II	2
30 minutes	III	3
45 minutes	IIII	4
60 minutes	~~IIII~~	5
75 minutes	~~IIII~~ II	7
90 minutes	III	3
105 minutes	I	1
120 minutes	IIII	4
135 minutes	I	1

Fig A.1

2 You should have designed a two-way table which looks like Fig A.2.

Activity \ Labourer	1	2	3	4	5
Lay a course of bricks 10 m long					
Mix a metre of concrete					
Dig two fence post holes					
Erect 5 m of scaffolding					

Fig A.2

7
Portable cassette player 12
Portable CD player 19
Mini hi-fi system 8
Midi hi-fi system 9
Car radio 4
Television 7
Video player 8
Satellite system 3
Washing machine 6
Tumble dryer 14
Refrigerator 6
Freezer 4
Vacuum cleaner 17

8 443.75 m

EXERCISE 7.1

1 (a) 21
(b) 24
(c) 32
(d) 50
(e) 90
(f) 13
(g) 26
(h) 12
(i) 34
(j) 8
(k) 8
(l) 12
(m) 36
(n) 9
(o) 9

2 £1100

3 348

4 (a) £387.75
(b) £537.94
(c) £1444.78
(d) £665.00
(e) £3035.47

5 7907 steps

6 (a) Machinist D makes 355 kites
(b) 1594 kites

EXERCISE 8.1

1 (a) 54
(b) 24
(c) 28
(d) 99
(e) 84
(f) 48
(g) 63
(h) 36
(i) 42
(j) 72
(k) 4
(l) 4
(m) 3
(n) 11
(o) 7

2 £28 144.13

3 (a) 9 lengths
(b) 44 lengths
(c) £170.37

4 (a) 448 bricks
(b) £67.20

5 $8 \times ((250 \times £18.75) + (450 \times £16.25) + (800 \times £13.50))$
$= 8 \times £22\,800 = £182\,400$

EXERCISE 9.1

1 (a) $\frac{7}{2}$

(b) $\frac{23}{5}$

(c) $\frac{21}{4}$

(d) $\frac{30}{11}$

(e) $\frac{25}{7}$

2 (a) $2\frac{3}{4}$

(b) $7\frac{1}{3}$

(c) $2\frac{3}{8}$

(d) $3\frac{2}{5}$

(e) $3\frac{5}{6}$

3 (a) $\frac{1}{4}$

(b) $\frac{1}{2}$

(c) $\frac{7}{9}$

(d) $\frac{11}{16}$

(e) $\frac{17}{22}$

4 (a) $\frac{8}{12} + \frac{3}{12} = \frac{11}{12}$

(b) $\frac{3}{6} + \frac{5}{6} = \frac{8}{6} = 1\frac{1}{3}$

(c) $\frac{15}{20} + \frac{16}{20} = \frac{31}{20} = 1\frac{11}{20}$

(d) $\frac{10}{30} + \frac{12}{30} + \frac{15}{30} = \frac{37}{30} = 1\frac{7}{30}$

(e) $\frac{6}{24} + \frac{4}{24} + \frac{16}{24} = \frac{26}{24} = 1\frac{2}{24} = 1\frac{1}{12}$

5 (a) $\frac{3}{8} - \frac{2}{8} = \frac{1}{8}$

(b) $\frac{11}{16} - \frac{10}{16} = \frac{1}{16}$

(c) $\frac{21}{35} - \frac{5}{35} = \frac{16}{35}$

(d) $\frac{15}{18} - \frac{8}{18} = \frac{7}{18}$

(e) $\frac{10}{16} - \frac{7}{16} = \frac{3}{16}$

6 (a) $\frac{4}{26} = \frac{2}{13}$

(b) $\frac{8}{63}$

(c) $\frac{4}{17} \times \frac{2}{1} = \frac{8}{17}$

(d) $\frac{1}{20} \times \frac{25}{3} = \frac{1}{4} \times \frac{5}{3} = \frac{5}{12}$

(e) $\frac{1}{22} \times \frac{11}{2} = \frac{1}{11} \times \frac{1}{2} = \frac{1}{22}$

7 (a) $\frac{3}{8} \times \frac{8}{5} = \frac{3}{5}$

(b) $\frac{7}{16} \times \frac{4}{1} = \frac{7}{4} = 1\frac{3}{4}$

(c) $\frac{13}{21} \times \frac{7}{3} = \frac{13}{3} \times \frac{1}{3} = \frac{13}{9} = 1\frac{4}{9}$

(d) $\frac{15}{18} \times \frac{12}{13} = \frac{15}{3} \times \frac{2}{13} = \frac{30}{39} = \frac{10}{13}$

(e) $\frac{4}{9} \times \frac{27}{24} = \frac{1}{1} \times \frac{3}{6} = \frac{3}{6} = \frac{1}{2}$

EXERCISE 9.2

1 (a) 1.63

(b) 231.879
(c) 47.67
(d) 5074.72
(e) 819.734

2 (a) 377.8
(b) 21.1
(c) 521.419
(d) 42465.4
(e) 39353.502

3 (a) 14.28
(b) 2601.36
(c) 17.9228
(d) 0.008544
(e) 24381.666

4 (a) 1.32
(b) 880
(c) 54
(d) 4.56
(e) 63.57

EXERCISE 9.3

1 (a) 271
(b) 76
(c) 320g
(d) 9045 m
(e) £174.30

2 (a) 546
(b) 66
(c) 720 kg
(d) 90 km
(e) 60p

3 (a) 20%
(b) 25%
(c) 12%
(d) 33%
(e) 44.5%

4 (a) 1000
(b) 7500
(c) £86
(d) 47584 km
(e) 9000 ml

EXERCISE 9.4

1 (a) Total parts $2 + 3 = 5$
Quantity per part $350 \div 5 = 70$
Multiply out
$2 \times 70 : 3 \times 70 = 140 : 210$
Answer = £140 : £210

(b) Total parts $13 + 12 = 25$
Quantity per part $1000 \div 25 = 40$
Multiply out
$13 \times 40 : 12 \times 40 = 520 : 480$
Answer = 520 km : 480 km

(c) Total parts $3 + 2 = 5$
Quantity per part $155 \div 5 = 31$
Multiply out $3 \times 31 : 2 \times 31 = 93 : 62$
Answer = 93 g : 62 g

(d) Total parts $19 + 11 = 30$
Quantity per part $660 \div 30 = 22$
Multiply out
$19 \times 22 : 11 \times 22 = 418 : 242$
Answer = 418 cm : 242 cm

(e) Total parts $27 + 20 = 47$
Quantity per part $282 \div 47 = 6$
Multiply out $27 \times 6 : 20 \times 6 = 162 : 120$
Answer = 162 ml : 120 ml

2 (a) Total parts $2 + 3 + 2 = 7$
Quantity per part $700 \div 7 = 100$
Multiply out
$2 \times 100 : 3 \times 100 : 2 \times 100$
$= 200 : 300 : 200$
Answer = 200 g : 300 g : 200 g

(b) Total parts $11 + 9 + 7 = 27$

Quantity per part $243 \div 27 = 9$

Multiply out

$$11 \times 9 : 9 \times 9 : 7 \times 9 = 99 : 81 : 63$$

Answer = 99 km : 81 km : 63 km

(c) Total parts $8 + 5 + 3 = 16$

Quantity per part $128 \div 16 = 8$

Multiply out

$$8 \times 8 : 5 \times 8 : 3 \times 8 = 64 : 40 : 24$$

Answer = 64 litres : 40 litres : 24 litres

(d) Total parts $2 + 1 + 3 = 6$

Quantity per part $5946 \div 6 = 991$

Multiply out

$$2 \times 991 : 1 \times 991 : 3 \times 991$$
$$= 1982 : 991 : 2973$$

Answer = £1982 : £991 : 2973

(e) Total parts $17 + 9 + 4 = 30$

Quantity per part $1.5 \div 30 = 0.05$

Multiply out

$$17 \times 0.05 : 9 \times 0.05 : 4 \times 0.05$$
$$= 0.85 : 0.45 : 0.2$$

Answer = 0.85 m : 0.45 m : 0.2 m

EXERCISE 9.5

1 Increase is £35 – £28 = £7

Percentage increase $= \frac{7}{28} \times 100 = 25$

Answer = 25%

2 $\frac{375}{600} = \frac{5}{8}$

3 Overspend is £10 700 – £10 000 = £700

Percentage overspend $= \frac{700}{10\,000} \times 100 = 7$

Answer = 7%

4 $\frac{375}{500} = \frac{75}{100} = \frac{3}{4}$

5 (a) $\frac{12}{76} \times 100 = 15.79$

Answer = 15.79%

(b) $\frac{54}{76} \times 100 = 71.05$

Answer = 71.05%

EXERCISE 9.6

1 $35 + \left(35 \times \frac{9}{100}\right) = 38.15$

Answer = £38.15

2 $\frac{6}{835} x = 90$

$x = 90 \times \frac{835}{6} = 12\,525$

Answer = £12 525

3 2.5 cm represents 5 km

1 cm represents $\frac{5}{2.5}$ km

9 cm represents $\frac{5}{2.5} \times 9$ km = 1.8 km

4 $365 \times \frac{12}{100} = 43.8$

Answer = £43.80

5 Total parts $13 + 11 + 8 = 32$

Quantity per part $1000 \div 32 = 31.25$

Multiply out

$$13 \times 31.25 : 11 \times 31.25 : 8 \times 31.25$$
$$= 406.25 : 343.75 : 250$$

6 $500 \times \frac{7}{10} = 350$

7 Size of extension = Size of new building – Size of old building

Size of new building = $37\ 500\ m^3 \times 3.5$
$= 131\ 250\ m^3$

Size of extension
$= 131\ 250\ m^3 - 37\ 500\ m^3$
$= 93\ 750\ m^3$

8 Brand A costs 35p for 230 g

$\frac{35}{230}$ p for 1 g

0.15p for 1 g

Brand B Costs 38p for 250 g

$\frac{38}{250}$ p for 1 g

0.15p for 1 g
Both brands are the same in terms of value for money.

9 $£\frac{24\ 120}{2} = £12\ 060$

$£\frac{24\ 120}{3} = £8\ 040$

$£\frac{24\ 120}{6} = £4\ 020$

£12 060 was spent on food, £8040 on clothes and £4020 on electricity

10 Total delivery
$= 552\ kg + \frac{8}{100} \times 552\ kg = 596.16\ kg$

EXERCISE 10.1

1 (a) 5.2 m
(b) 680 mm
(c) 9 cm
(d) 37 m
(e) 27 m

2 $2 \times 5\ m + 2 \times 3\ m + 2 \times 0.5\ m$
$= 10\ m + 6\ m + 1\ m = 17\ m$

3 $2 \times 2206\ m + 2 \times 3673\ m$
$= 11\ 758\ m$

4 2 circuits

5 1200 mm + 900 mm + 900 mm = 3000 mm

EXERCISE 10.2

(a) $16\ cm^2$
(b) $27\ cm^2$
(c) $12\ cm^2$
(d) $18\ cm^2$
(e) $22\ cm^2$
(f) $17\ cm^2$

EXERCISE 10.3

1 $35\ m^2$

2 $1856\ cm^2 = 0.1856\ m^2$

3 $106\ 470\ m^2$

4 $\frac{1}{2} \times 600\ mm \times 1200\ mm = 360\ 000\ mm^2$
$= 3600\ cm^2$
$= 0.36\ m^2$

5 $\frac{1}{2} \times 60\ cm \times 70\ cm = 2100\ cm^2$

6 $45\ m \times 30\ m = 1350\ m^2$ (Yes)

7 Blood bank requires $6 \times 5\ m^2 = 30\ m^2$
Trellis Hall is $5\ m \times 9\ m = 45\ m^2$
Cumberland Hall is $7\ m \times 8\ m = 56\ m^2$

Trellis Hall is more suitable.

8 Parking bay of ferry is $25.2\ m \times 153\ m$
$= 3855.6\ m^2$
$\frac{3855.6}{16.2} = 238$

9 Area of base = $\frac{1}{2} \times 7.5 \text{ cm} \times 6.5 \text{ cm}$

$= 24.375 \text{ cm}^2$

195 cm^2 allocated by shop for x boxes of 24.375 cm^2

$x = \frac{195}{24.375} = 8$

10 Each sign is $\frac{1}{2} \times 0.6 \text{ m} \times 0.6 \text{ m} = 0.18 \text{ m}^2$

Using a 2 m^2 sheet of metal, x signs of 0.18 m^2 can be made

$x = \frac{2}{0.18} = 11$

EXERCISE 10.4

1 $20 \text{ mm} \times 20 \text{ mm} \times 20 \text{ mm} = 8000 \text{ mm}^3$

2 $3.5 \text{ m} \times 4.3 \text{ m} \times 11.2 \text{ m} = 168.56 \text{ m}^3$

3 $47 \text{ cm} \times 62 \text{ cm} \times 132 \text{ cm} = 384\,648 \text{ cm}^3$
$= 0.385 \text{ m}^3$

4 $3000 \text{ mm} \times 2550 \text{ mm}^2 = 7\,650\,000 \text{ mm}^3$
$= 7650 \text{ cm}^3$
$= 0.0077 \text{ m}^3$

5 $\frac{1}{2} \times 30 \text{ cm} \times 17 \text{ cm} \times 5 \text{ cm} = 1275 \text{ cm}^3$

6 Old cabinets volume:
$5 \times 50 \text{ cm} \times 60 \text{ cm} \times 140 \text{ cm}$
$= 2\,100\,000 \text{cm}^3$
New cabinets volume:
$3 \times 110 \text{ cm} \times 50 \text{ cm} \times 200 \text{ cm}$
$= 3\,300\,000 \text{ cm}^3$
Additional space:
$3\,300\,000 \text{ cm}^3 - 2\,100\,000 \text{ cm}^3$
$= 1\,200\,000 \text{ cm}^3$

7 $320 \times 4.6 \text{ m} \times 3.2 \text{ m} \times 8.7 \text{ m} = 40\,980.48 \text{ m}^3$

8 Each charm has a volume of:
$4 \text{ mm} \times 4.5 \text{ mm}^2 = 18 \text{ mm}^3$
There is 2430 mm^3 available material to make x charms of 18 mm^3

$x = \frac{2430}{18} = 135$

9 $30 \text{ cm} \times 32 \text{ cm} \times 57 \text{ cm}$ contains 14 books with volume x
$54\,720 \text{cm}^3$ contains 14 books with volume x

$x = \frac{54\,720}{14} = 3908 \text{ cm}^3$

10 Volume of stone:
$60 \text{ cm} \times 60 \text{ cm} \times 90 \text{ cm} = 324\,000 \text{ cm}^3$

Volume of support:
$= \frac{1}{2} \times 60 \text{ cm} \times 60 \text{ cm} \times 90 \text{ cm}$
$= 162\,000 \text{ cm}^3$

Volume wasted:
$324\,000 \text{ cm}^3 - 162\,000 \text{ cm}^3 = 162\,000 \text{ cm}^3$

EXERCISE 11.1

1 $3.142 \times 30 \text{ mm} = 94.25 \text{ mm}$

2 $3.142 \times 5^2 = 78.54$
Area $= 78.54 \text{ cm}^2$

3 Circumference $= 3.142 \times 65 \text{ mm} \times 2$
$= 408.41 \text{ mm}$
Area $= 3.142 \times 65^2 \text{ mm}$
$= 13\,273.23 \text{ mm}^2$

4 $3.142 \times r^2 = 50.272 \text{m}^2$
$r^2 = \frac{50.272}{3.142}$
$= 16 \text{ m}^2$
$r = \sqrt{16 \text{ m}^2}$
$= 4 \text{m}$

5 $3.142 \times d = 9.426\text{cm}$

$$d = \frac{9.426}{3.142}\text{ cm}$$

$$= 3\text{ cm}$$

6 $3.142 \times \left(\frac{57}{2}\right)^2 = 2551.76$

(Answer is 2551 cm^2)

7 $3.142 \times 2.5\text{ m} = 7.85\text{ m}$

8 $r = 75$ cm

Circumference $= 3.142 \times 75 \times 2$

$= 471.3$ cm

Area $= 3.142 \times 75^2$ cm

$= 17\,673.75\text{ cm}^2$

9 $3.142 \times d = 47\,130\text{mm}$

$$d = \frac{47\,130}{3.142}\text{ mm}$$

$$= 15\,000\text{ mm}$$

10 $3.142 \times r^2 = 153\text{ m}^2$

$$r^2 = \frac{153}{3.142}\text{ m}^2 = 49\text{ m}^2$$

$$r = \sqrt{49}\text{ m} = 7\text{m}$$

EXERCISE 12.1

1 £10 000 + (£0.35 × 500) = £10 175

2 £505 600 – £17 850 = £487 750

3 564 pairs + 75 pairs = 639 pairs

4 £13.50 + (£0.568 × 580) = £342.94

5 $\frac{£2350}{27\,500} = £0.09$

= 9p

6 £980 – £205 – £55 = £720

EXERCISE 13.1

1 (a) 7
(b) 5
(c) 2
(d) 8
(e) 60

2 £109.99

3 3 hours

4 £2.75

5 8 miles/hour

6 £35 + (£1.27 × 158) = £235.66

7 Let x = price of sand, therefore $3x$ = price of cement

$$60x + (10 \times 3x) = 90$$
$$60x + 30x = 90$$
$$90x = 90$$
$$x = 1$$

Sand = £1 per bag, cement = £3 per bag

8 Let x = cost of the spreadsheet package, therefore $x + 21.5$ = cost of database package

$$15(x + 21.5) + 21x = 1632$$
$$15x + 322.5 + 21x = 1632$$
$$36x + 322.5 = 1632$$
$$36x = 1632 - 322.5$$
$$= 1309.5$$
$$x = 36.375$$

Spreadsheet package costs £36.38.

9 Let x = number of hard chairs

$$15x + 20(30 - x) = 500$$
$$15x + 600 - 20x = 500$$
$$600 - 5x = 500$$
$$600 - 500 = 5x$$
$$100 = 5x$$
$$x = 20$$

They buy 20 hard chairs.

10 (a) Let x = cost of the badminton court

$10x = 42.5$

$x = 4.25$

(b) Let y = cost of the basketball court, given that the badminton court costs £4.25, then

$y + (6 \times 4.25) = 42.5$

$y + 25.5 = 42.5$

$y = 17$

Check:

$(2 \times 4.25) + (2 \times 17) = 42.5$

EXERCISE 14.1

1 See Fig A.3.

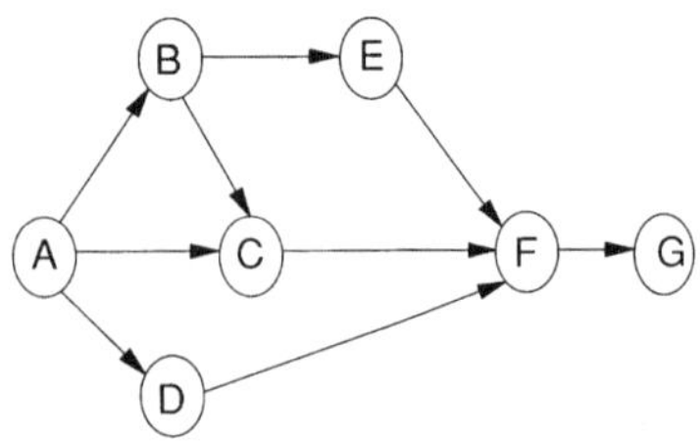

Fig A.3

2 See Fig A.4.

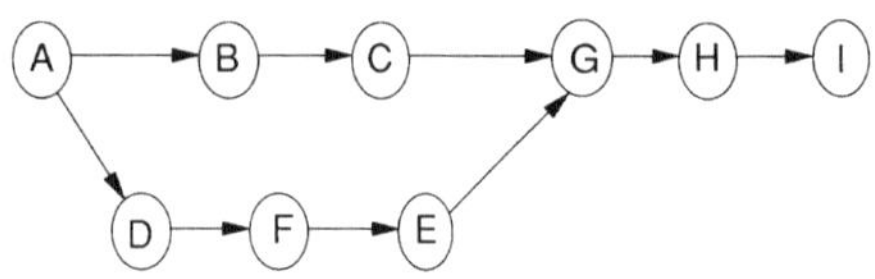

Fig A.4

3 See Figs A.5, A.6.

Activity	Prior Activity	Activity	Prior Activity
A	—	G	C
B	A	H	G
C	A	I	H
D	A	J	D,F,I
E	B	K	J
F	E	L	K

Fig A.5

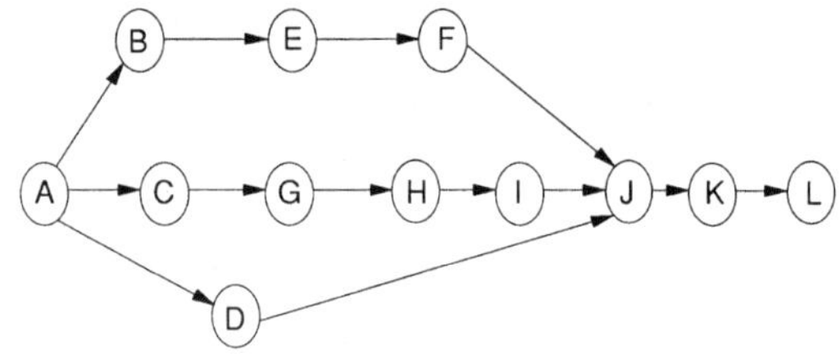

Fig A.6

EXERCISE 15.1

1 $3.142 \times 33^2\ \text{mm}^2 \times 130\text{mm} = 444\,812.94\ \text{mm}^3$

2 Area of concrete $= 10\ \text{m} \times 25\ \text{m}$
$= 250\ \text{m}^2$

Area of grass $= (10\ \text{m} - 1.5\ \text{m} - 1.5\ \text{m}) \times (25\ \text{m} - 1.5\ \text{m} - 1.5\ \text{m})$
$= 7\ \text{m} \times 22\ \text{m}$
$= 154\ \text{m}^2$

Area of pavement = Area of concrete – Area of grass

$250\ \text{m}^2 - 154\ \text{m}^2 = 96\ \text{m}^2$

3 $3.142 \times \left(\frac{51}{2}\right)^2 - 3.142 \times \left(\frac{48}{2}\right)^2 = 233.26$

Cross-sectional area = $233.26\ \text{mm}^2$

4 $3.142 \times \left(\frac{180}{2}\right)^2 \times 92 = 2341418.4$

Volume = $2341418.4\ \text{cm}^3$
$= 2.34\ \text{m}^3$

5 $(65 \times 24) + 2(65 \times 120) + 2(24 \times 120)$
$= 22\,920$

$22\,902\ \text{cm}^2 = 2.29\ \text{m}^2$

6 Area of wood $= 1.5\ \text{m}^2$
$= 15\,000\ \text{cm}^2$

Area of table top $= 3.142 \times \left(\frac{55}{2}\right)^2$ cm^2

$= 2376.14$ cm^2

Area of support $= 2(50 \times 65)$ cm^2

$= 6500$ cm^2

Area of waste = Area of wood – Area of table top – Area of support

$= 6123.86$ cm^2

7 Area of football pitch $= 110$ m $\times 75$ m

$= 8250$ m^2

Area of roller $= 3.142 \times 1.2$m $\times 1.7$m

$= 6.41$ m^2

No. of revolutions $= \dfrac{\text{Area of football pitch}}{\text{Area of roller}}$

$= 1287$

EXERCISE 16.1

1
$$\begin{aligned}(x + 6x) + (5y - 5y) &= 13 + 8\\ 7x &= 21\\ x &= 3\\ 3 + 5y &= 13\\ 5y &= 10\\ y &= 2\end{aligned}$$

2
$$\begin{aligned}4x + 12y &= 64\\ (4x - 4x) + (12y - 3y) &= 64 - 37\\ 9y &= 27\\ y &= 3\\ 2x + 18 &= 32\\ 2x &= 14\\ x &= 7\end{aligned}$$

3
$$\begin{aligned}12x + 3y &= 144\\ (12x + 3x) + (3y - 3y) &= 144 + 6\\ 15x &= 150\\ x &= 10\\ 40 + y &= 48\\ y &= 8\end{aligned}$$

4 $3x + 2y = 79$ (1)

$2x + 5y = 115$ (2)

$$\begin{aligned}(1)\times 2 \qquad 6x + 4y &= 158\\ (2)\times 3 \qquad 6x + 15y &= 345\\ (6x - 6x) + (15y - 4y) &= 345 - 158\\ 11y &= 187\\ y &= 17\\ 3x + 34 &= 79\\ 3x &= 45\\ x &= 15\end{aligned}$$

5 $7x - 5y = 8$ (1)

$2x + 7y = 36$ (2)

$$\begin{aligned}(1)\times 7 \qquad 49x - 35y &= 56\\ (2)\times 5 \qquad 10x + 35y &= 180\\ (10x + 49x) + (35y - 35y) &= 180 + 56\\ 59x &= 236\\ x &= 4\\ 8 + 7y &= 36\\ 7y &= 28\\ y &= 4\end{aligned}$$

EXERCISE 16.2

1 Let the number of £5 notes be x and the number of £10 notes be y

$$\begin{aligned}x + y &= 24\\ 5x + 10y &= 195\\ 10x + 10y &= 240\\ (10x - 5x) + (10y - 10y) &= 240 - 195\\ 5x &= 45\\ x &= 9\end{aligned}$$

Answer: 9 £5 notes

2 Machine A $x + 2y = 16$

Machine B $3x + 2y = 36$

$$\begin{aligned}(3x - x) + (2y - 2y) &= 36 + 6\\ 2x &= 20\\ x &= 10\\ 10 + 2y &= 16\\ 2y &= 6\\ y &= 3\end{aligned}$$

Answer: 10 of product x and 3 of product y

3 Let the number of dresses be x and the number of shirts be y

$$\begin{aligned} x + y &= 500 \\ 30x + 22y &= 12\,264 \\ 30x + 30y &= 15\,000 \\ (30x - 30x) + (30y - 22y) &= 15\,000 - 12\,264 \\ 8y &= 2736 \\ y &= 342 \\ x + 342 &= 500 \\ x &= 158 \end{aligned}$$

Answer: 158 dresses and 342 shirts were made

4 Peter: $38x + 4y = 352$ (1)
Matthew: $40x + 6y = 392$ (2)

$$\begin{aligned} (1) \times 3 \quad 114x + 12y &= 1056 \\ (2) \times 2 \quad 80x + 12y &= 784 \\ (114x - 80x) + (12y - 12y) &= 1056 - 784 \\ 34x &= 272 \\ x &= 8 \\ 304 + 4y &= 352 \\ 4y &= 48 \\ y &= 12 \end{aligned}$$

Answer: basic pay is £8 per hour and overtime is £12 per hour

EXERCISE 16.3

1 $4x - 2x > 9 - 5$
$2x > 4$
$x > 2$

2 $6x + x \geqslant 18 - 11$
$7x \geqslant 7$
$x \geqslant 1$

3 $-2x - 3x > 7 - 22$
$-5x > -15$
$x < 3$

4 $3x - 2x \geqslant -5 - 19$
$x \geqslant -24$

5 $-6x < -9 - 25$
$-6x < -34$
$x > \frac{-34}{-6}$
$x > 5\frac{2}{3}$

6 Let the number of bandages be x and the number of pressure pads be y

$$\begin{aligned} 2x + 5y &< 120 \\ 2x + (5 \times 14) &< 120 \\ 2x + 70 &< 120 \\ 2x &< 50 \\ x &< 25 \end{aligned}$$

Answer: less than 25 bandages

7

$$\begin{aligned} x + 2y &\leqslant 64 \\ 3x + y &\leqslant 72 \\ 3x + 6y &\leqslant 192 \\ (3x - 3x) + (6y - y) &\leqslant 192 - 72 \\ 5y &\leqslant 120 \\ y &\leqslant 24 \\ x + 48 &\leqslant 64 \\ x &\leqslant 16 \end{aligned}$$

Answer: 16 widgets and 24 gadgets

EXERCISE 17.1

1 $£3500(1 + 0.09)^3 = £4532.60$

2 $£1500(1 + 0.12)^4 = £2360.28$

3 $£15\,500(1 + 0.055)^{12} = £29\,468.72$

4 $700(1 + 0.11)^2 = 862.47$
$862.47 - 700 = 162.47$
Interest = £162.47

5 $950\,(1 + 0.08)^7 = 1628.13$
$1628.13 - 950 = 678.13$
Interest = £678.13

EXERCISE 17.2

1 $£5500(1 - 0.2)^5 = £1802.24$

2 $£15\,000(1 - 0.12)^5 = £7915.98$

3 $£30\,000\,(1 - 0.14)^{10} = £6639.05$

4
$$300 = P(1 - 0.08)^5$$
$$300 = P \times 0.659\,081\,5$$
$$455.18 = P$$
Original value: £455.18

5
$$5000 = P(1 - 0.16)^5$$
$$5000 = P \times 0.418\,211\,9$$
$$11955.66 = P$$
Original cost: £11955.66

EXERCISE 19.1

1 (a) cube
(b) cuboid
(c) triangular-based pyramid
(d) cylinder
(e) prism
(f) square-based pyramid

2 See Fig A.7 (a)–(e) (not to scale).

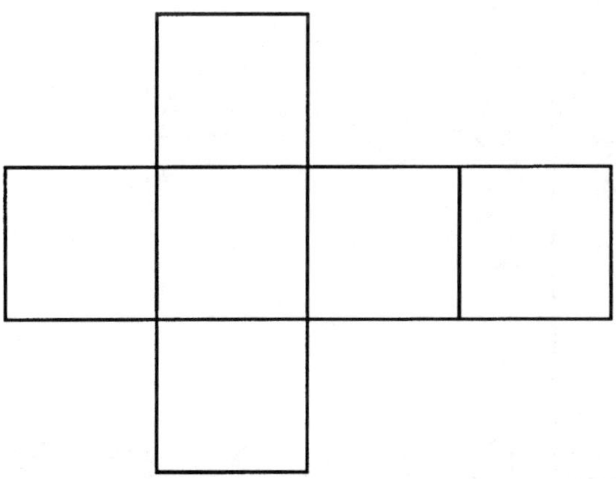

Fig A.7(a)

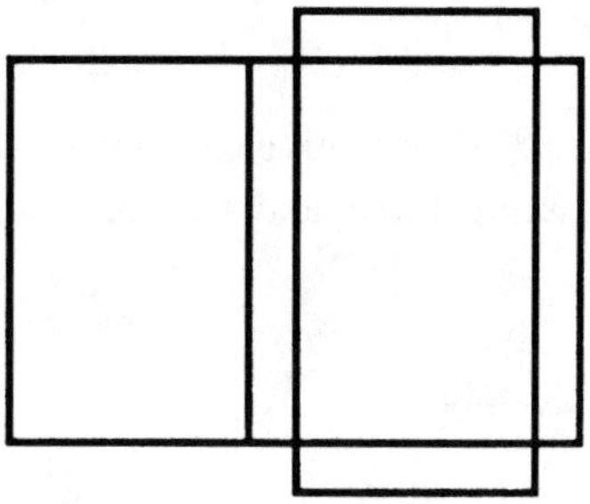

Fig A.7(b)

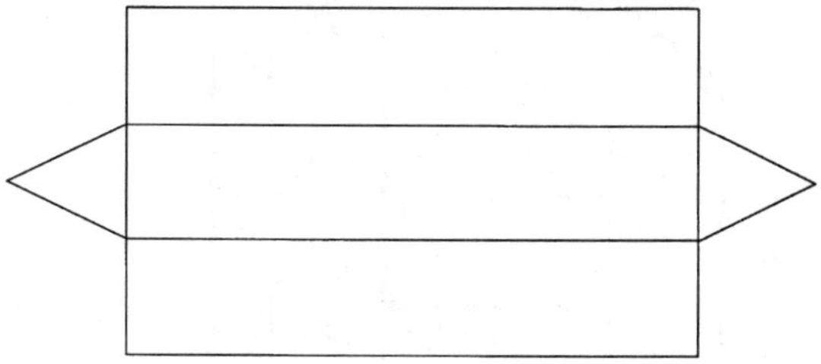

Fig A.7(c)

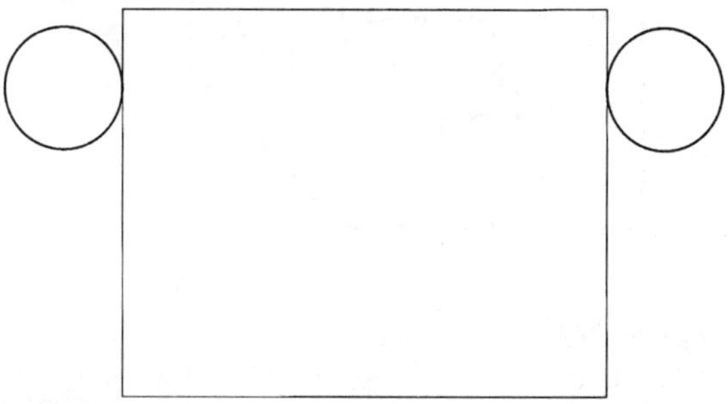

Fig A.7(d)

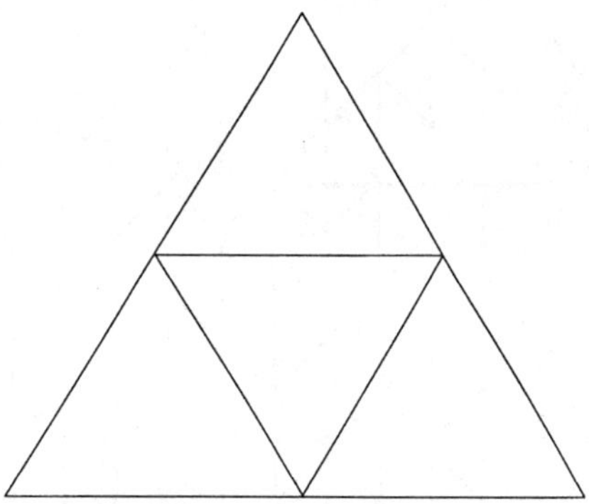

Fig A.7(e)

EXERCISE 19.2

1 The objects are a 3-drawer filing cabinet, a can, a square-based pyramid and a carton (milk/juice).

2 See Fig A.8 (a), (b), (c).

Fig A.8(a)

Fig A.8(b)

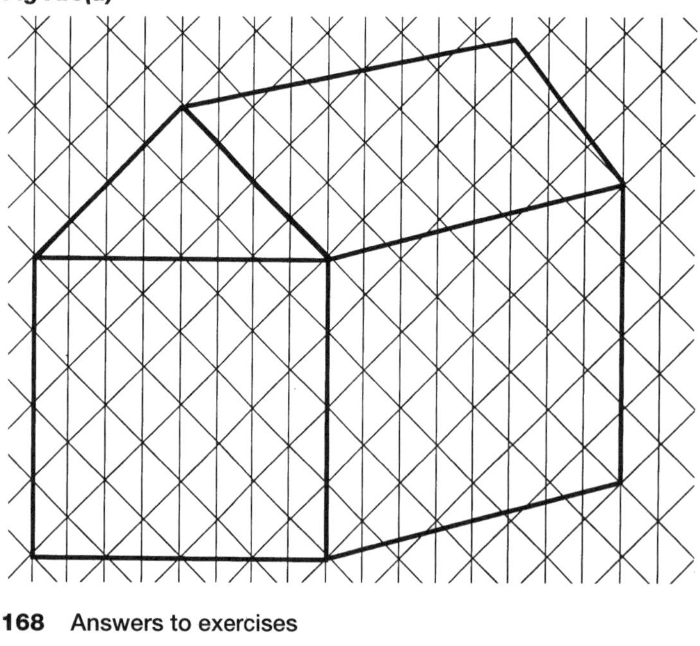

Fig A.8(c)

EXERCISE 19.3

1 lounge

2 1

3 bathroom

4 kitchen

5 bedroom 2

EXERCISE 20.1

1 Your graph should resemble the one in Fig A.9.

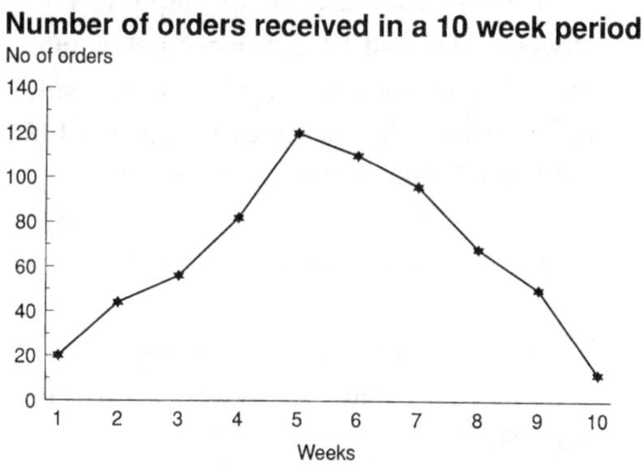

Fig A.9

2 Your bar chart should look like the one in Fig A.10.

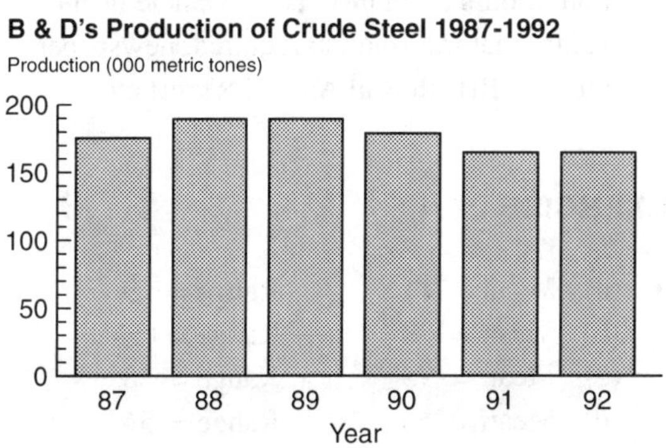

Fig A.10

3 You should have constructed a bar chart and line graph which look like Fig A.11 (a) and (b).

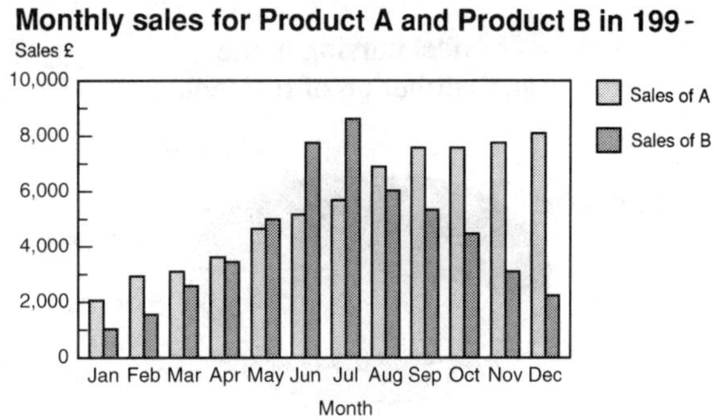

Fig A.11(a)

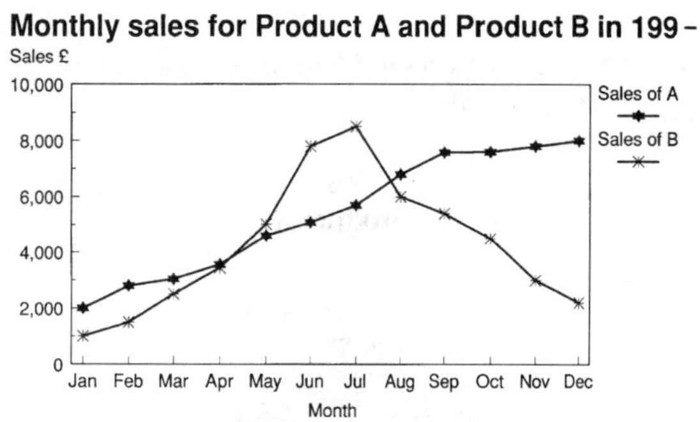

Fig A.11(b)

EXERCISE 20.2

1 Your scatter graph should look like the one in Fig A.12.

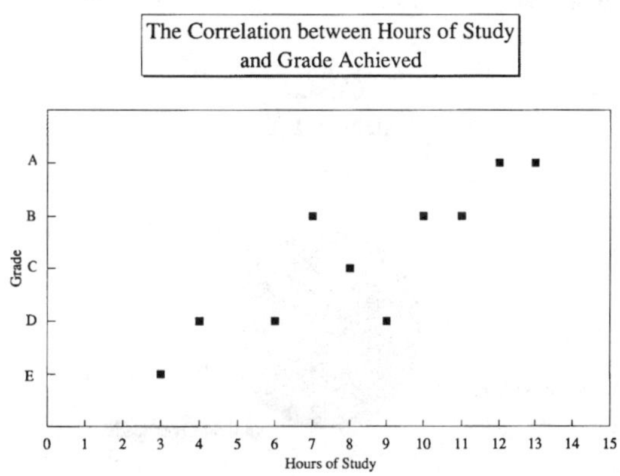

Fig A.12

EXERCISE 20.3

1 You should have constructed a pie chart as in Fig A.13.

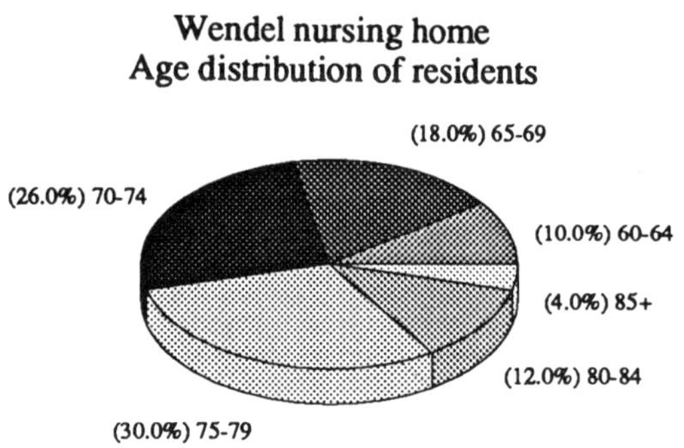

Fig A.13

2 You pie charts should look like those in Fig A.14 (a) and (b).

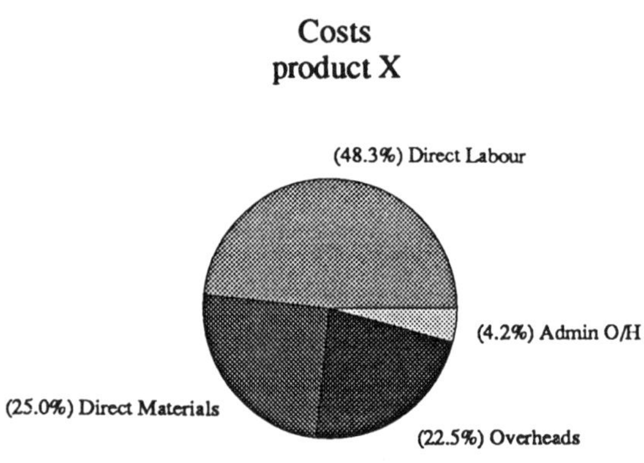

Fig A.14(a)

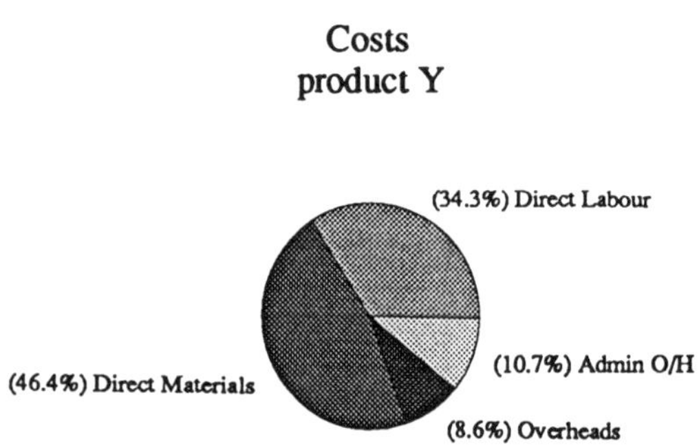

Fig A.14(b)

EXERCISE 20.4

1 1991

2 1988

3 1992

4 £7.8 million

5 Industry: metal; Injury: major

6 Consume most: apples (marginally greater than other fresh fruit)
Least popular: bananas

7 Private sector construction increased output in 1986, and peaked in 1988. Its output levels between 1989 and 1992 dropped below the output level achieved in 1988, but were still higher than the levels achieved between 1982 and 1986. Public sector output dropped in 1986. It rose marginally in 1987 and 1988, but didn't achieve its highest level until 1992.

8 Sea travel has remained fairly constant. Air travel has increased steadily between 1981 and 1992.

9 Passengers

10 Possible causes could be major train accidents or an increase in rail travel caused by a reduction in fairs or an increase in vehicle licences. Additional information required: newspaper articles, British Rail Annual Report etc.

EXERCISE 22.1

1 (a) Mean = 11   Range = 9
(b) Mean = 100   Range = 151
(c) Mean = 95   Range = 8
(d) Mean = 258   Range = 56

2

	Prod A	*Prod B*
Mean	5380	4233
Range	6000	7500

Product B because it has a lower average and a greater range.

3

	Machine A	*Machine B*
Mean	499	500
Range	9	15

Conclusions: Machine A is more consistent because it has a lower range, but its mean is lower than the required weight. Machine B has the required mean weight, but has a greater range and so is less consistent. Draw your own conclusions!

EXERCISE 23.1

1 $\frac{1}{20} = 0.$

2 (a) $\frac{1}{6} = 0.17$

(b) $\frac{3}{6} = 0.5$

(c) $\frac{2}{6} = 0.33$

3 (a) $\frac{13}{25} = 0.52$

(b) $\frac{12}{25} \times \frac{13}{25} = \frac{156}{625} = 0.2496$

4 (a) $\frac{4}{20} = 0.2$

(b) $\frac{10}{20} \times \frac{10}{20} = \frac{100}{400} = 0.25$

5 (a) $\frac{12}{40} = 0.3$

(b) $\frac{8}{40} = 0.2$

(c) $\frac{5}{40} + \frac{3}{40} + \frac{1}{40} + \frac{1}{40} = \frac{10}{40} = 0.25$

(d) $\frac{10}{40} \times \frac{10}{40} = \frac{100}{1600} = 0.0625$

EXERCISE 24.1

1 $\frac{2}{200} = \frac{1}{100} = 0.01$

2 $\frac{54}{450} = \frac{6}{50} = 0.12$

3 $\frac{54}{1000} = \frac{27}{500} = 0.054$

4 Product A: $\frac{349}{500} = 0.698$

Product B: $\frac{296}{428} = 0.692$

Product A is more efficient.

5 $\frac{2}{40} = \frac{1}{20} = 0.05$